T0224338

# Computer Techniques for Dynamic Modeling of DC-DC Power Converters

# Synthesis Lectures on Power Electronics

Editor

**Jerry Hudgins,** *University of Nebraska*

*Synthesis Lectures on Power Electronics* will publish 50- to 100-page publications on topics related to power electronics, ancillary components, packaging and integration, electric machines and their drive systems, as well as related subjects such as EMI and power quality. Each lecture develops a particular topic with the requisite introductory material and progresses to more advanced subject matter such that a comprehensive body of knowledge is encompassed. Simulation and modeling techniques and examples are included where applicable. The authors selected to write the lectures are leading experts on each subject who have extensive backgrounds in the theory, design, and implementation of power electronics, and electric machines and drives.

The series is designed to meet the demands of modern engineers, technologists, and engineering managers who face the increased electrification and proliferation of power processing systems into all aspects of electrical engineering applications and must learn to design, incorporate, or maintain these systems.

Computer Techniques for Dynamic Modeling of DC-DC Power Converters
Farzin Asadi
2018

Robust Control of DC-DC Converters: The Kharitonov's Approach with MATLAB® Codes
Farzin Asadi
2018

Dynamics and Control of DC-DC Converters
Farzin Asadi and Kei Eguchi
2018

Analysis of Sub-synchronous Resonance (SSR) in Doubly-fed Induction Generator (DFIG)-Based Wind Farms
Hossein Ali Mohammadpour and Enrico Santi
2015

Power Electronics for Photovoltaic Power Systems
Mahinda Vilathgamuwa, Dulika Nayanasiri, and Shantha Gamini
2015

Digital Control in Power Electronics, 2nd Edition
Simone Buso and Paolo Mattavelli
2015

Transient Electro-Thermal Modeling of Bipolar Power Semiconductor Devices
Tanya Kirilova Gachovska, Bin Du, Jerry L. Hudgins, and Enrico Santi
2013

Modeling Bipolar Power Semiconductor Devices
Tanya K. Gachovska, Jerry L. Hudgins, Enrico Santi, Angus Bryant, and Patrick R. Palmer
2013

Signal Processing for Solar Array Monitoring, Fault Detection, and Optimization
Mahesh Banavar, Henry Braun, Santoshi Tejasri Buddha, Venkatachalam Krishnan, Andreas
Spanias, Shinichi Takada, Toru Takehara, Cihan Tepedelenlioglu, and Ted Yeider
2012

The Smart Grid: Adapting the Power System to New Challenges
Math H.J. Bollen
2011

Digital Control in Power Electronics
Simone Buso and Paolo Mattavelli
2006

Power Electronics for Modern Wind Turbines
Frede Blaabjerg and Zhe Chen
2006

Computer Techniques for Dynamic Modeling of DC-DC Power Converters

Farzin Asadi

ISBN: 978-3-031-01376-8    paperback
ISBN: 978-3-031-02504-4    ebook
ISBN: 978-3-031-00325-7    hardcover

DOI 10.1007/978-3-031-02504-4

A Publication in the Springer series
*SYNTHESIS LECTURES ON POWER ELECTRONICS*

Lecture #12
Series ISSN
Print 1931-9525    Electronic 1931-9533

# Computer Techniques for Dynamic Modeling of DC-DC Power Converters

Farzin Asadi
Kocaeli University, Kocaeli, Turkey

*SYNTHESIS LECTURES ON POWER ELECTRONICS #12*

# ABSTRACT

Computers play an important role in the analyzing and designing of modern DC-DC power converters. This book shows how the widely used analysis techniques of averaging and linearization can be applied to DC-DC converters with the aid of computers. Obtained dynamical equations may then be used for control design.

The book is composed of two chapters. Chapter 1 focuses on the extraction of control-to-output transfer function. A second-order converter (a buck converter) and a fourth-order converter (a Zeta converter) are studied as illustrative examples in this chapter. Both ready-to-use software packages, such as PLECS® and MATLAB® programming, are used throught this chapter.

The input/output characteristics of DC-DC converters are the object of considerations in Chapter 2. Calculation of input/output impedance is done with the aid of MATLAB® programming in this chapter. The buck, buck-boost, and boost converter are the most popular types of DC-DC converters and used as illustrative examples in this chapter.

This book can be a good reference for researchers involved in DC-DC converters dynamics and control.

# KEYWORDS

control of DC-DC converters, dynamics of DC-DC converters, input impedance of DC-DC converters, MATLAB®, PLECS®, small-signal model, state space averaging (SSA), modeling of power electronics converters, output impedance of DC-DC converters, system identification

*I dedicate this book to my parents and my lovely family.*

# Contents

# Acknowledgments

The author gratefully acknowledges the MathWorks® support for this project.


Farzin Asadi
August 2018

CHAPTER 1

# Extraction of Small-Signal Transfer Functions Using PLECS® and MATLAB®

## 1.1 INTRODUCTION

Switch-mode converters are widely used today to provide power processing for applications ranging from computing and communications to medical electronics, appliance control, transportation, and high-power transmission. Their high efficiency, small size, low weight, and reduced cost make them a good alternative for conventional linear power supplies, even at low power levels.

Switched DC-DC converters are nonlinear variable structure systems. The nonlinearities arise primarily due to switching, power devices, and passive components, such as inductors. Various techniques can be found in the literature to obtain a Linear Time Invariant (LTI) model of a switched DC-DC converter. The most well-known methods are [Ericson and Maksimovic, 2001]: Circuit averaging (CA) and State Space Averaging (SSA).

CA replaces the semiconductor switches (nonlinear part of the converter) with an (averaged) equivalent linear circuit. In this method manipulations are carried out based on a circuit diagram [Akbarabadi et al., 2013].

SSA uses the duty cycle as a weighting factor and combines the state equations into a single averaged state equation. The procedure of state space averaging is explained in detail in Asadi and Eguchi [2018] and Suntio [2009].

SSA has a number of advantages over circuit averaging technique. These include:

- the ability to obtain more transfer functions than was possible using circuit averaging technique; and

- the ability to more easily obtain both DC and AC transfer functions.

Foundation of SSA was laid down in Middlebrook and Cuk [1977] and later extended in Tymerski and Vorperian [1986], Chen and Ngo [2001], Sun et al. [2001], as well as many other publications. The first attempt to model Discontinuous Conduction Mode (DCM) is presented in Cuk and Middlebrook [1977]. A unified SSA-based method to develop both Continuous Current Mode (CCM) and DCM was developed by Suntio [2006]. A comprehensive survey of the modeling issues can be found in Maksimovic et al. [2001].

This chapter presents a tutorial exposition of modeling DC-DC converters using computer tools. Extraction of converters dynamics (except for the simple second-order converters) using pencil-and-paper is a difficult and error-prone task.

Both ready-to-use software packages (PLECS®) and MATLAB® programming are used to show the process of DC-DC converter modeling.

This chapter is organized as follows. Extraction of small-signal transfer functions using PLECS® is studied in the second section. PLECS® only gives the Bode plot of the converter; it does not provide the algebraic transfer function of the converter. Therefore, MATLAB® (and system identification toolbox) is used to fit an algebraic transfer function to the Bode plot calculated by PLECS® in the third section. MATLAB® implementation of SSA is studied in the fourth section and a Zeta converter is studied as an illustrative example there.

## 1.2    EXTRACTION OF SMALL-SIGNAL TRANSFER FUNCTIONS USING PLECS®

Piecewise Linear Electrical Circuit Simulation (PLECS®) is a software tool for system-level simulation of electrical circuits developed by Plexim [Allmeling and Hammer, 1999]. PLECS® comes in two versions: standalone and Simulink. The standalone has its own solver and can be run independently. The Simulink version, as the name suggests, runs under the MATLAB®/Simulink environment and uses the Simulink solver. We use the standalone version in this chapter. PLECS® has a free trial version which can be used for period of one month.

PLECS® can be used to extract the small-signal transfer functions of DC-DC converters. To obtain the small-signal transfer functions, an user must add the "Small Signal Perturbation" and "Small Signal Response" blocks (see Fig. 1.1) to the suitable places of schematic.

We extract the small signal transfer functions of a Buck converter (see Fig. 1.2) with the following components values:

$$Vs = 10, \; L = 100 \; \mu\text{H}, \; C = 100 \; \mu\text{F}, \; rC = 0.01 \; \Omega, \; R = 5 \; \Omega, \; F_{switching} = 50 \; \text{KHz}.$$

The schematic diagrams shown in Figs. 1.3, 1.4, and 1.5 are used to obtain the frequency response of control-to-output $\left( \frac{\tilde{v}_o(j\omega)}{\tilde{d}(j\omega)} \right)$, audio susceptibility $\left( \frac{\tilde{v}_o(j\omega)}{\tilde{v}_g(j\omega)} \right)$, and output impedance $\left( \frac{\tilde{v}_o(j\omega)}{\tilde{i}_o(j\omega)} \right)$, respectively. The seetings of "Symmetrical PWM" block is shown in Fig. 1.6.

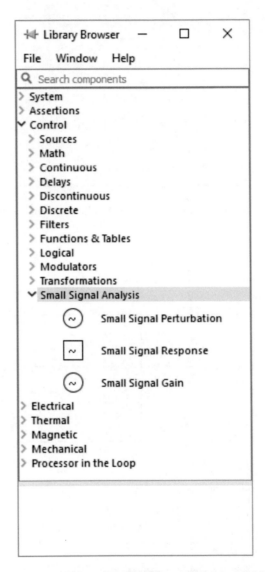

Figure 1.1: "Small Signal Perturbation" and "Small-Signal Response" blocks. Use the "Search components" text box to find the desired compoent if you don't know its place.

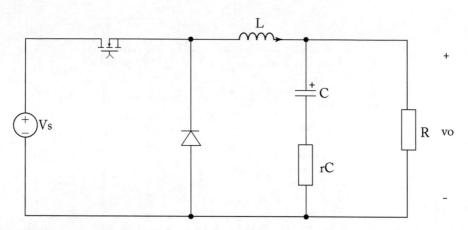

Figure 1.2: Schematic of a buck converter. MOSFET and diode are assumed to be ideal.

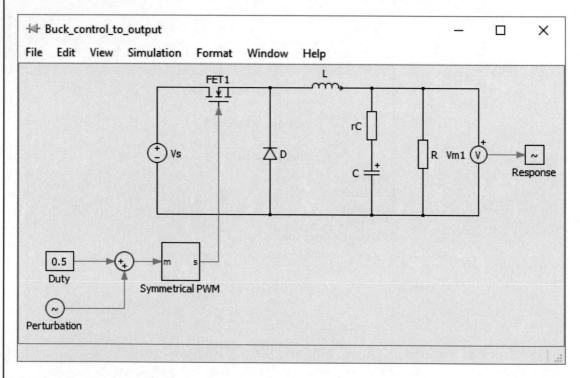

Figure 1.3: Schematic diagram to obtain the frequency response of control-to-output $\left(\frac{\tilde{v}_o(j\omega)}{\tilde{d}(j\omega)}\right)$.

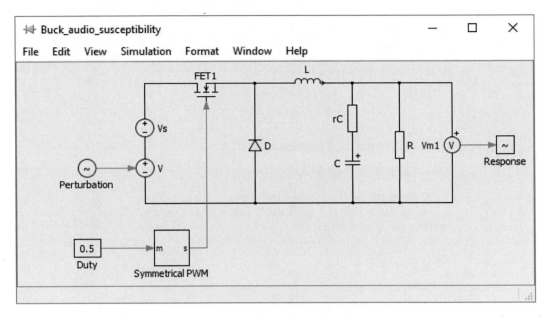

Figure 1.4: Schematic diagram to obtain the frequency response of audio susceptibility $\left(\frac{\tilde{v}_o(j\omega)}{\tilde{v}_s(j\omega)}\right)$. Block named "V" is a "Voltage Source (Controlled)" block (see Fig. 1.7).

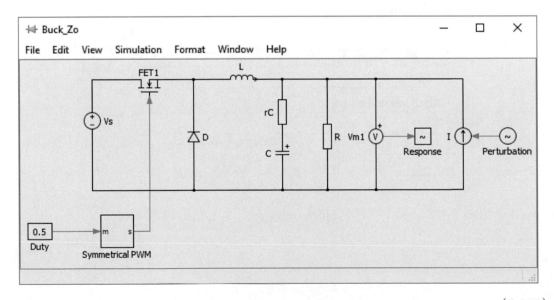

Figure 1.5: Schematic diagram to obtain the frequency response of output impedance $\left(\frac{\tilde{v}_o(j\omega)}{\tilde{i}_o(j\omega)}\right)$. Block named "I" is a "Current Source (Controlled)" block (see Fig. 1.8).

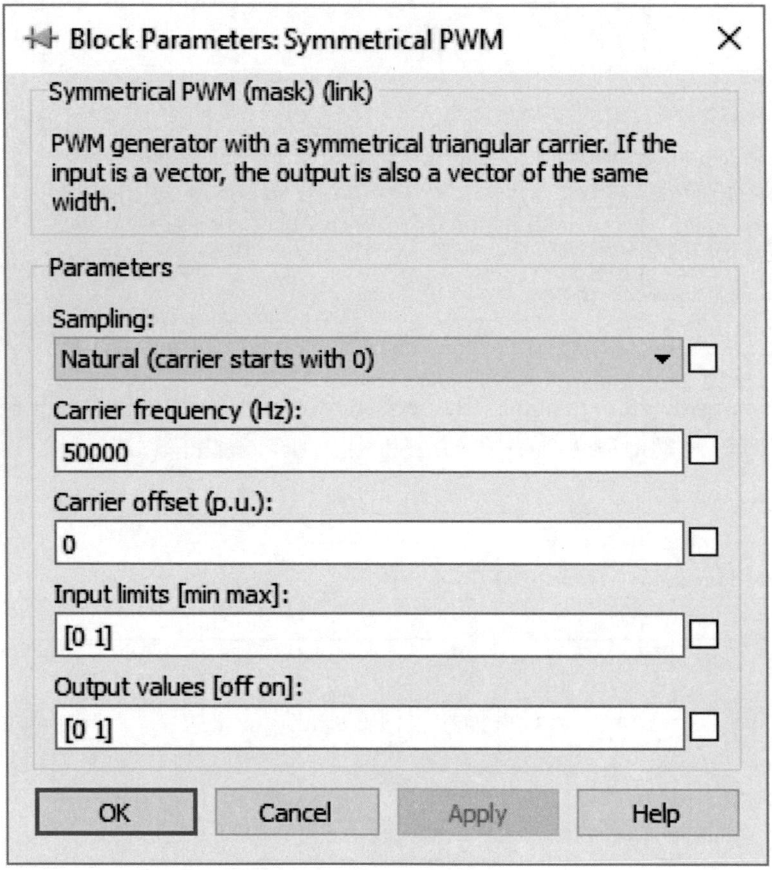

Figure 1.6: The "Symmetrical PWM" block (see Figs. 1.3, 1.4, and 1.5) settings.

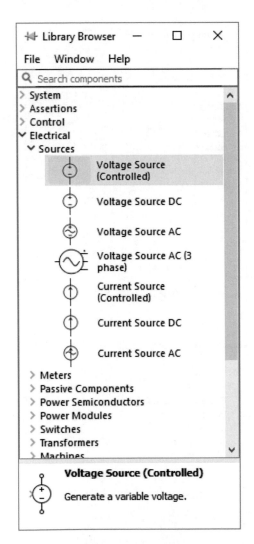

Figure 1.7: "Voltage Source (Controlled)" block.

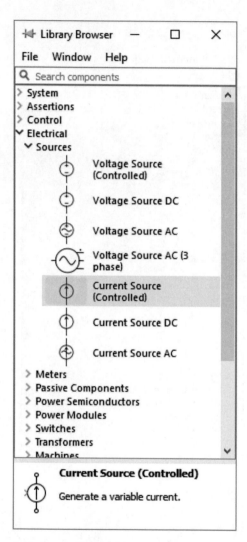

Figure 1.8: "Current Source (Controlled)" block.

In Fig. 1.9, "Analysis tools…" is clicked to obtain the converter dynamics.

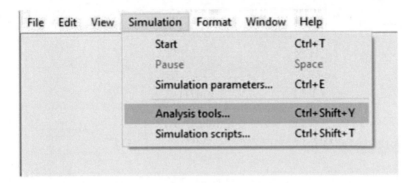

Figure 1.9: Converter dynamics are obtained with the aid of "Analysis tools" choice.

After clicking the "Analysis tools…", the window shown in Fig. 1.10 will appear.

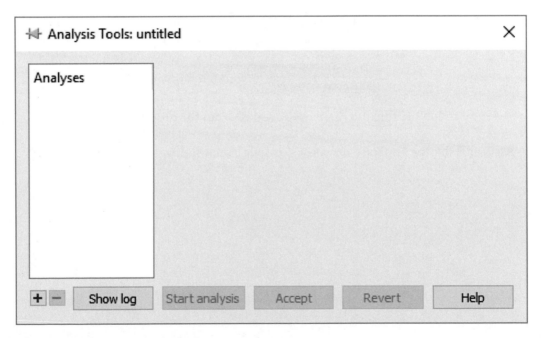

Figure 1.10: "Analysis Tools" window.

Click the + button. The window shown in Fig. 1.11 will appear.

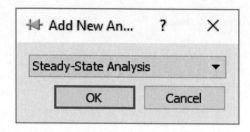

Figure 1.11: "Add New Analysis" window.

Select the "Impulse Response Analysis" from the drop down list.

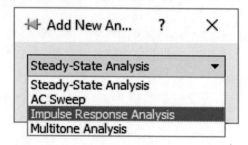

Figure 1.12: Selection of "Impulse Response Analysis" from the drop down list.

After clicking the "Impulse Response Analysis", the window shown in Fig. 1.13 will appear.

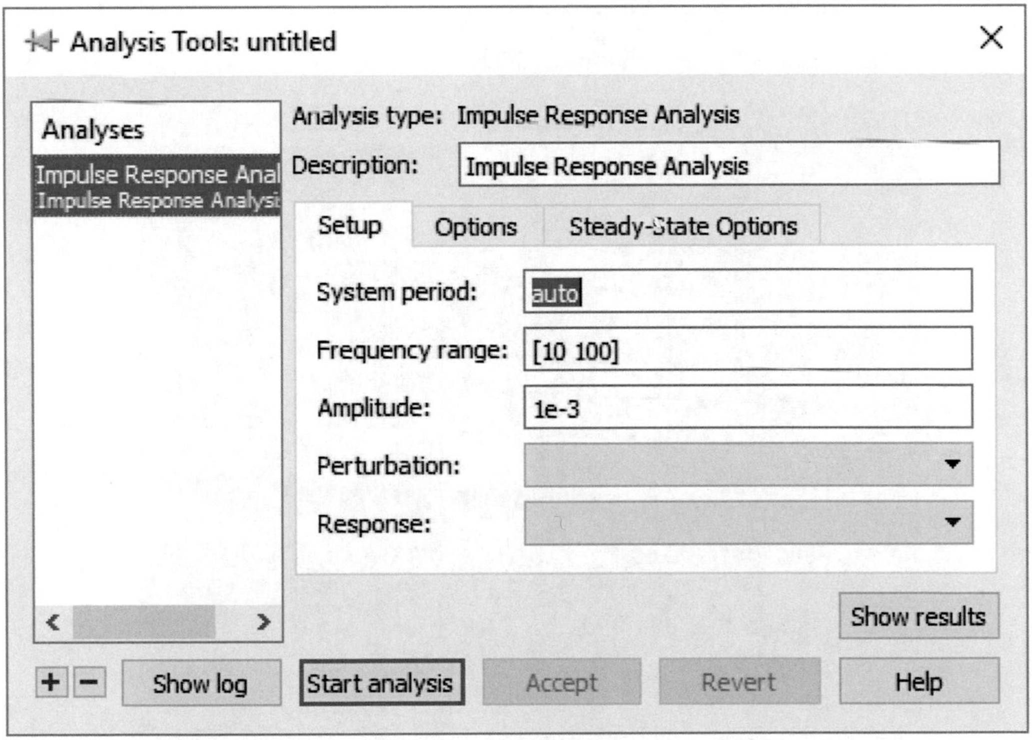

Figure 1.13: "Impulse Response Analysis" settings.

You can enter the desired band of frequencies in the "Frequency range" box. PLECS®
takes the desired band of frequencies in Hertz. The "Perturbation" and "Response" boxes must
be filled with the names given to small-signal perturbation and small-signal response block,
respectively.

Click the "Start analysis" button to obtain the converter dynamics. Simulation results are
shown in Figs. 1.14, 1.15, and 1.16. The control-to-output frequency response can be used to
design the controller.

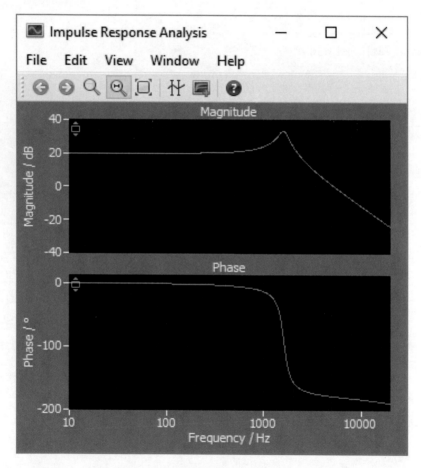

Figure 1.14: Frequency response of control-to-output $\left(\frac{\tilde{v}_o(j\omega)}{\tilde{d}(j\omega)}\right)$.

PLECS® does not provide the algebraic transfer functions. It only draws the Bode diagrams. The next section introduces amethod to convert the obtained graph into an algebraic
transfer function.

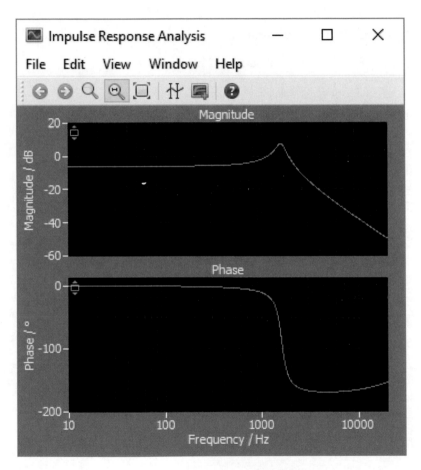

Figure 1.15: Frequency response of audio susceptibility $\left(\frac{\tilde{v}_o(j\omega)}{\tilde{v}_g(j\omega)}\right)$.

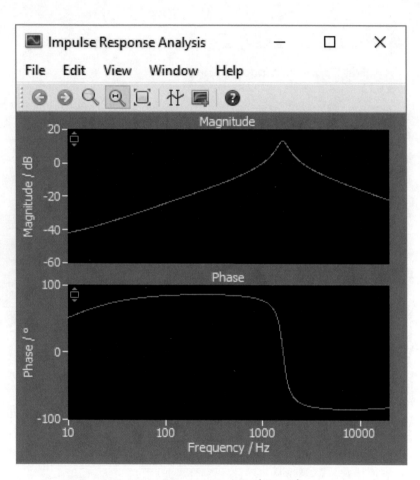

Figure 1.16: Frequency response of output impedance $\left(\frac{\tilde{v}_o(j\omega)}{\tilde{i}_o(j\omega)}\right)$.

## 1.3    OBTAINING THE SUITABLE ALGEBRAIC TRANSFER FUNCTIONS FOR THE GRAPH

PLECS® can export the obtained graphs as Camma Seperated Values (CSV) files (Fig. 1.17). MATLAB® can read CSV files. The code shown below can be used to fit a transfer function for the obtained graph.

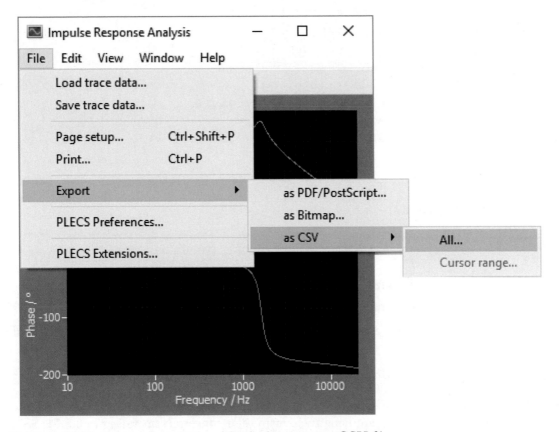

Figure 1.17: Exporting the calculated Bode diagram as a CSV file.

```
%This program fits a transfer function to the
%obtained frequency response.
%Use Notepad to clear the first line of data.csv
%before using the ``csvread'' function.

data=csvread('data.csv');
w=2*pi*data(:,1);
```

```
val=10.^(data(:,2)/20).*exp(j*data(:,3)*pi/180);
sys=frd(val,w);
bode(sys),grid on
hold on


%This section lets you click on the drawn bode
%diagram. A trafnsfer is fitted based on the
%selected points. You can use the MATLAB's ``tfest''
%function as well.
NumberOfDesiredPoints=20;
[freq, resp_db]=ginput(NumberOfDesiredPoints);
for i=1:NumberOfDesiredPoints
    resp(i)=10^(resp_db(i)/20);
end
sys=frd(resp,freq);
ord=2;
W=fitmagfrd(sys,ord);
Wtf=tf(W);
bode(Wtf,'r')
```

Figure 1.18 shows the fitted transfer function and the PLECS® analysis result on the same plot. (We estimated a transfer function for the control-to-output $\left(\frac{\tilde{v}_o(s)}{\tilde{d}(s)}\right)$ transfer function.) As shown, the estimated transfer function lies on the PLECS® results. This shows the goodness of fit.

## 1.4   DYNAMICS OF CCM CONVERTERS

In this section we study the dynamics of DC-DC converters operating in CCM. SSA is one of the most important tools for studying the dynamics of converters operating in CCM. SSA has two important steps: averaging and linearization. The SSA procedure can be summarized as follows [Asadi and Eguchi, 2018].

1. Circuit differential equations are written for different working modes (i.e., on/off state of semiconductor switches).

2. Equations are time averaged over one period.

3. Steady-state operating points are calculated by equating the derivative terms to zero.

4. The averaged equations are linearized around the steady-state operating point found in the third step.

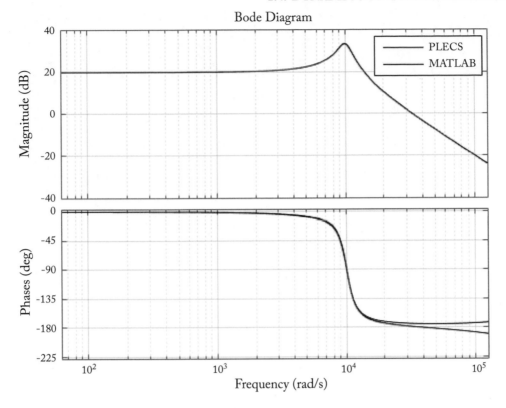

Figure 1.18: Bode diagram of estimated transfer function $\left(1000\frac{s+10^6}{s^2+2100s+10^8}\right)$ vs. result provided by PLECS®.

Applying this procedure is quite tedious and error prone for pencil-and-paper analysis (especially if the converter order is high). MATLAB® can be very helpful to do the mathematical machinery of SSA. We show the usefulness of MATLAB® to extract the converter small-signal transfer functions with an example.

### 1.4.1 DYNAMICS OF A ZETA CONVERTER

Schematic of a Zeta converter is shown in Fig. 1.19. The Zeta converter is composed of two switches: a MOSFET switch and a diode. In this schematic, $Vg, rg, Li, rLi, Ci, rCi$, and $R$ shows input DC source, input DC source internal resistance, $i$th inductor, $i$th inductor Equivalent Series Resistance (ESR), $i$th capacitor, $i$th capacitor ESR, and load, respectively. $iO$ is a fictitious current source added to the schematic in order to calculate the output impedance of converter. In this section we assume that converter works in CCM. MOSFET switch is controlled with the aid of a Pulse Width Modulator (PWM) controller. MOSFET switch keeps closed for

$D.T$ seconds and $(1 - D).T$ seconds open. $D$ and $T$ show duty ratio and switching period, respectively.

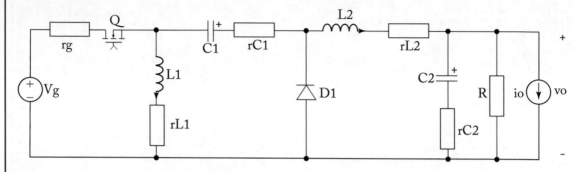

Figure 1.19: Schematic of Zeta converter.

When MOSFET is closed, the diode is opened (Fig. 1.20).

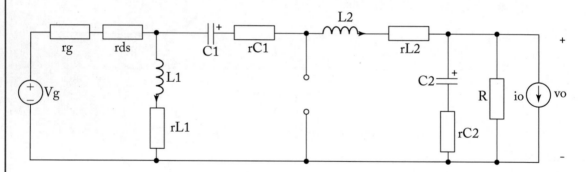

Figure 1.20: Closed MOSFET.

The circuit differential equations can be written as:

$$L_1 \frac{di_{L_1}}{dt} = -\left(r_{L_1} + r_g + r_{ds}\right) i_{L_1} - \left(r_g + r_{ds}\right) i_{L_2} + v_g \tag{1.1}$$

$$L_2 \frac{di_{L_2}}{dt} = -\left(r_g + r_{ds}\right) i_{L_1} - \left(r_g + r_{ds} + r_{C_1} + r_{L_2} + \frac{R \times r_{C_2}}{R + r_{C_2}}\right) i_{L_2}$$

$$+ v_{C_1} - \frac{R}{R + r_{C_2}} v_{C_2} + \frac{R \times r_{C_2}}{R + r_{C_2}} i_o + v_g \tag{1.2}$$

$$C_1 \frac{dv_{C_1}}{dt} = -i_{L_2} \tag{1.3}$$

$$C_2 \frac{dv_{C_2}}{dt} = \frac{R}{R + r_{C_2}} i_{L_2} - \frac{1}{R + r_{C_2}} v_{C_2} - \frac{R}{R + r_{C_2}} i_o \tag{1.4}$$

$$v_o = r_{C_2} C_2 \frac{dv_{C_2}}{dt} + v_{C_2} = \frac{R \times r_{C_2}}{R + r_{C_2}} i_{L_2} + \frac{R}{R + r_{C_2}} v_{C_2} - \frac{R \times r_{C_2}}{R + r_{C_2}} i_o. \tag{1.5}$$

When MOSFET is opened, the diode is closed (Fig. 1.21). In Fig. 1.21, forward-biased diode is modeled with a voltage source (VD) and a series resistance (rD).

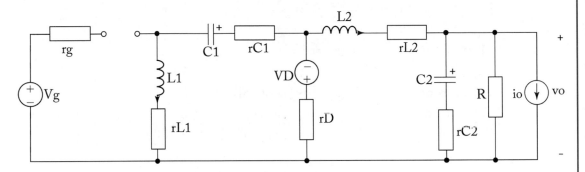

Figure 1.21: Opened MOSFET.

The circuit differential equations can be written as:

$$L_1 \frac{di_{L_1}}{dt} = -\left(r_{L_1} + r_{C_1} + r_D\right) i_{L_1} - r_D i_{L_2} - v_{C_1} - v_D \tag{1.6}$$

$$L_2 \frac{di_{L_2}}{dt} = -r_D i_{L_1} - \left(r_D + r_{L_2} + \frac{R \times r_{C_2}}{R + r_{C_2}}\right) i_{L_2} - \frac{R}{R + r_{C_2}} v_{C_2} + \frac{R \times r_{C_2}}{R + r_{C_2}} i_o - v_D \tag{1.7}$$

$$C_1 \frac{dv_{C_1}}{dt} = i_{L_1} \tag{1.8}$$

$$C_2 \frac{dv_{C_2}}{dt} = \frac{R}{R + r_{C_2}} i_{L_2} - \frac{1}{R + r_{C_2}} v_{C_2} - \frac{R}{R + r_{C_2}} i_o \tag{1.9}$$

$$v_o = r_{C_2} C_2 \frac{dv_{C_2}}{dt} + v_{C_2} = \frac{R \times r_{C_2}}{R + r_{C_2}} i_{L_2} + \frac{R}{R + r_{C_2}} v_{C_2} - \frac{R \times r_{C_2}}{R + r_{C_2}} i_o. \tag{1.10}$$

Consider a converter with the values given in Table 1.1.

The following program extracts the converter transfer functions. It uses the SSA to extract the converter transfer functions.

Table 1.1: The Zeta converter parameters (see Fig. 1.10)

| | Nominal Value |
|---|---|
| Output voltage, Vo | 5.2 V |
| Duty ratio, D | 0.23 |
| Input DC source voltage, Vg | 20 V |
| Input DC source internal resistance, rg | 0.0 Ω |
| MOSFET drain-source resistance, rds | 10 mΩ |
| Capacitor, C1 | 100 μF |
| Capacitor Equivaluent Series Resistance (ESR), rC1 | 0.19 Ω |
| Capacitor, C2 | 220 μF |
| Capacitor Equivaluent Series Resistance (ESR), rC2 | 0.095 Ω |
| Inductor, L1 | 100 μH |
| Inductor ESR, rL1 | 1 mΩ |
| Inductor, L2 | 55 μH |
| Inductor ESR, rL2 | 0.55 mΩ |
| Diode voltage drop, vD | 0.7 V |
| Diode forward resistance, rD | 10 mΩ |
| Load resistor, R | 6 Ω |
| Switching Frequency, Fsw | 100 KHz |

```
%This program calculates the small signal
%transfer functions of Zeta converter
clc
clear all

VG=20;       %Average value of input DC source
rg=0;        %Internal resistance of input DC source
rds=.01;     %MOSFET on resistance
C1=100e-6;   %Capacitor C1 value
C2=220e-6;   %Capacitor C2 value
rC1=.19;     %Capacitor C1 Equivalent Series Resistance(ESR)
rC2=.095;    %Capacitor C2 Equivalent Series Resistance(ESR)
L1=100e-6;   %Inductor L1 value
L2=55e-6;    %Inductor L2 value
```

```
rL1=1e-3;    %Inductor L1 Equivalent Series Resistance(ESR)
rL2=.55e-3;  %Inductor L2 Equivalent Series Resistance(ESR)
rD=.01;      %Diode series resistance
VD=.7;       %Diode voltage drop
R=6;         %Load resistance
D=.23;       %Duty cylcle
IO=0;        %Average value of output current source
fsw=100e3;   %Switching frequency

syms iL1 iL2 vC1 vC2 io vg vD d
%iL1: Inductor L1 current
%iL2: Inductor L2 current
%vC1: Capacitor C1 voltage
%vC2: Capacitor C2 voltage
%io : Output current source
%vg : Input DC source
%vD : Diode voltage drop
%d  : Duty cycle

%Closed MOSFET Equations
diL1_dt_MOSFET_close=(-(rL1+rg+rds)*iL1-(rg+rds)*iL2+vg)/L1;
diL2_dt_MOSFET_close=(-(rg+rds)*iL1-(rg+rds+rC1+rL2+R*
    rC2/(R+rC2))*iL2+vC1-R/(R+rC2)*vC2+R*rC2/(R+rC2)*io+vg)/L2;
dvC1_dt_MOSFET_close=(-iL2)/C1;
dvC2_dt_MOSFET_close=(R/(R+rC2)*iL2-1/(R+rC2)*vC2-R/(R+rC2)*
    io)/C2;
vo_MOSFET_close=R*rC2/(R+rC2)*iL2+R/(R+rC2)*vC2-R*rC2/(R+rC2)*io;

%Opened MOSFET Equations
diL1_dt_MOSFET_open=(-(rL1+rC1+rD)*iL1-rD*iL2-vC1-vD)/L1;
diL2_dt_MOSFET_open=(-rD*iL1-(rD+rL2+R*rC2/(R+rC2))*
    iL2-R/(R+rC2)*vC2+R*rC2/(R+rC2)*io-vD)/L2;
dvC1_dt_MOSFET_open=(iL1)/C1;
dvC2_dt_MOSFET_open=(R/(R+rC2)*iL2-1/(R+rC2)*
    vC2-R/(R+rC2)*io)/C2;
vo_MOSFET_open=R*rC2/(R+rC2)*iL2+R/(R+rC2)*vC2-R*rC2/(R+rC2)*io;

%Averaging
averaged_diL1_dt=simplify(d*diL1_dt_MOSFET_close+(1-d)*
```

```
    diL1_dt_MOSFET_open);
averaged_diL2_dt=simplify(d*diL2_dt_MOSFET_close+(1-d)*
    diL2_dt_MOSFET_open);
averaged_dvC1_dt=simplify(d*dvC1_dt_MOSFET_close+(1-d)*
    dvC1_dt_MOSFET_open);
averaged_dvC2_dt=simplify(d*dvC2_dt_MOSFET_close+(1-d)*
    dvC2_dt_MOSFET_open);
averaged_vo=simplify(d*vo_MOSFET_close+(1-d)*vo_MOSFET_open);

%Substituting the steady values of input
%DC voltage source, Diode voltage drop,
%Duty cycle and output current source
%and calculating the DC operating point
right_side_of_averaged_diL1_dt=subs(averaged_diL1_dt,
    [vg vD d io],[VG VD D IO]);
right_side_of_averaged_diL2_dt=subs(averaged_diL2_dt,
    [vg vD d io],[VG VD D IO]);
right_side_of_averaged_dvC1_dt=subs(averaged_dvC1_dt,
    [vg vD d io],[VG VD D IO]);
right_side_of_averaged_dvC2_dt=subs(averaged_dvC2_dt,
    [vg vD d io],[VG VD D IO]);

DC_OPERATING_POINT=
solve(right_side_of_averaged_diL1_dt==0,
    right_side_of_averaged_diL2_dt==0,
    right_side_of_averaged_dvC1_dt==0,
    right_side_of_averaged_dvC2_dt==0,'iL1','iL2','vC1','vC2');

IL1=eval(DC_OPERATING_POINT.iL1);
IL2=eval(DC_OPERATING_POINT.iL2);
VC1=eval(DC_OPERATING_POINT.vC1);
VC2=eval(DC_OPERATING_POINT.vC2);
VO=eval(subs(averaged_vo,[iL1 iL2 vC1 vC2 io],
    [IL1 IL2 VC1 VC2 IO]));

disp('Operating point of converter')
disp('----------------------------')
disp('IL1(A)=')
disp(IL1)
```

```
disp('IL2(A)=')
disp(IL2)
disp('VC1(V)=')
disp(VC1)
disp('VC2(V)=')
disp(VC2)
disp('VO(V)=')
disp(VO)
disp('---------------------------')

%Linearizing the averaged equations around
%the DC operating point. We want to obtain
%the matrix A, B, C, and D.
%      x=Ax+Bu
%      y=Cx+Du
%
%where,
%      x=[iL1 iL2 vC1 vC2]'
%      u=[io vg d]'
%Since we used the variables D for steady
%state duty ratio and C to show the capacitors
%values we use AA, BB, CC, and DD instead of A,
%B, C and D.

%Calculating the matrix A
A11=subs(simplify(diff(averaged_diL1_dt,iL1)),
    [iL1 iL2 vC1 vC2 d io],[IL1 IL2 VC1 VC2 D IO]);
A12=subs(simplify(diff(averaged_diL1_dt,iL2)),
    [iL1 iL2 vC1 vC2 d io],[IL1 IL2 VC1 VC2 D IO]);
A13=subs(simplify(diff(averaged_diL1_dt,vC1)),
    [iL1 iL2 vC1 vC2 d io],[IL1 IL2 VC1 VC2 D IO]);
A14=subs(simplify(diff(averaged_diL1_dt,vC2)),
    [iL1 iL2 vC1 vC2 d io],[IL1 IL2 VC1 VC2 D IO]);

A21=subs(simplify(diff(averaged_diL2_dt,iL1)),
    [iL1 iL2 vC1 vC2 d io],[IL1 IL2 VC1 VC2 D IO]);
A22=subs(simplify(diff(averaged_diL2_dt,iL2)),
    [iL1 iL2 vC1 vC2 d io],[IL1 IL2 VC1 VC2 D IO]);
A23=subs(simplify(diff(averaged_diL2_dt,vC1)),
```

```
      [iL1 iL2 vC1 vC2 d io],[IL1 IL2 VC1 VC2 D IO]);
A24=subs(simplify(diff(averaged_diL2_dt,vC2)),
      [iL1 iL2 vC1 vC2 d io],[IL1 IL2 VC1 VC2 D IO]);


A31=subs(simplify(diff(averaged_dvC1_dt,iL1)),
      [iL1 iL2 vC1 vC2 d io],[IL1 IL2 VC1 VC2 D IO]);
A32=subs(simplify(diff(averaged_dvC1_dt,iL2)),
      [iL1 iL2 vC1 vC2 d io],[IL1 IL2 VC1 VC2 D IO]);
A33=subs(simplify(diff(averaged_dvC1_dt,vC1)),
      [iL1 iL2 vC1 vC2 d io],[IL1 IL2 VC1 VC2 D IO]);
A34=subs(simplify(diff(averaged_dvC1_dt,vC2)),
      [iL1 iL2 vC1 vC2 d io],[IL1 IL2 VC1 VC2 D IO]);


A41=subs(simplify(diff(averaged_dvC2_dt,iL1)),
      [iL1 iL2 vC1 vC2 d io],[IL1 IL2 VC1 VC2 D IO]);
A42=subs(simplify(diff(averaged_dvC2_dt,iL2)),
      [iL1 iL2 vC1 vC2 d io],[IL1 IL2 VC1 VC2 D IO]);
A43=subs(simplify(diff(averaged_dvC2_dt,vC1)),
      [iL1 iL2 vC1 vC2 d io],[IL1 IL2 VC1 VC2 D IO]);
A44=subs(simplify(diff(averaged_dvC2_dt,vC2)),
      [iL1 iL2 vC1 vC2 d io],[IL1 IL2 VC1 VC2 D IO]);


AA=eval([A11 A12 A13 A14;
         A21 A22 A23 A24;
         A31 A32 A33 A34;
         A41 A42 A43 A44]);


%Calculating the matrix B
B11=subs(simplify(diff(averaged_diL1_dt,io)),
      [iL1 iL2 vC1 vC2 d vD io vg],
      [IL1 IL2 VC1 VC2 D VD IO VG]);
B12=subs(simplify(diff(averaged_diL1_dt,vg)),
      [iL1 iL2 vC1 vC2 d vD io vg],
      [IL1 IL2 VC1 VC2 D VD IO VG]);
B13=subs(simplify(diff(averaged_diL1_dt,d)),
      [iL1 iL2 vC1 vC2 d vD io vg],
      [IL1 IL2 VC1 VC2 D VD IO VG]);


B21=subs(simplify(diff(averaged_diL2_dt,io)),
```

```
   [iL1 iL2 vC1 vC2 d vD io vg],
   [IL1 IL2 VC1 VC2 D VD IO VG]);
B22=subs(simplify(diff(averaged_diL2_dt,vg)),
   [iL1 iL2 vC1 vC2 d vD io vg],
   [IL1 IL2 VC1 VC2 D VD IO VG]);
B23=subs(simplify(diff(averaged_diL2_dt,d)),
   [iL1 iL2 vC1 vC2 d vD io vg],
   [IL1 IL2 VC1 VC2 D VD IO VG]);


B31=subs(simplify(diff(averaged_dvC1_dt,io)),
   [iL1 iL2 vC1 vC2 d vD io vg],
   [IL1 IL2 VC1 VC2 D VD IO VG]);
B32=subs(simplify(diff(averaged_dvC1_dt,vg)),
   [iL1 iL2 vC1 vC2 d vD io vg],
   [IL1 IL2 VC1 VC2 D VD IO VG]);
B33=subs(simplify(diff(averaged_dvC1_dt,d)),
   [iL1 iL2 vC1 vC2 d vD io vg],
   [IL1 IL2 VC1 VC2 D VD IO VG]);


B41=subs(simplify(diff(averaged_dvC2_dt,io)),
   [iL1 iL2 vC1 vC2 d vD io vg],
   [IL1 IL2 VC1 VC2 D VD IO VG]);
B42=subs(simplify(diff(averaged_dvC2_dt,vg)),
   [iL1 iL2 vC1 vC2 d vD io vg],
   [IL1 IL2 VC1 VC2 D VD IO VG]);
B43=subs(simplify(diff(averaged_dvC2_dt,d)),
   [iL1 iL2 vC1 vC2 d vD io vg],
   [IL1 IL2 VC1 VC2 D VD IO VG]);


BB=eval([B11 B12 B13;
         B21 B22 B23;
         B31 B32 B33;
         B41 B42 B43]);

%Calculating the matrix C
C11=subs(simplify(diff(averaged_vo,iL1)),
   [iL1 iL2 vC1 vC2 d io],
   [IL1 IL2 VC1 VC2 D IO]);
C12=subs(simplify(diff(averaged_vo,iL2)),
```

```
      [iL1 iL2 vC1 vC2 d io],
      [IL1 IL2 VC1 VC2 D IO]);
  C13=subs(simplify(diff(averaged_vo,vC1)),
      [iL1 iL2 vC1 vC2 d io],
      [IL1 IL2 VC1 VC2 D IO]);
  C14=subs(simplify(diff(averaged_vo,vC2)),
      [iL1 iL2 vC1 vC2 d io],
      [IL1 IL2 VC1 VC2 D IO]);

  CC=eval([C11 C12 C13 C14]);

  D11=subs(simplify(diff(averaged_vo,io)),
      [iL1 iL2 vC1 vC2 d vD io vg],
      [IL1 IL2 VC1 VC2 D VD IO VG]);
  D12=subs(simplify(diff(averaged_vo,vg)),
      [iL1 iL2 vC1 vC2 d vD io vg],
      [IL1 IL2 VC1 VC2 D VD IO VG]);
  D13=subs(simplify(diff(averaged_vo,d)),
      [iL1 iL2 vC1 vC2 d vD io vg],
      [IL1 IL2 VC1 VC2 D VD IO VG]);

  %Calculating the matrix D
  DD=eval([D11 D12 D13]);

  %Producing the State Space Model and obtaining
  %the small signal transfer functions
  sys=ss(AA,BB,CC,DD);
  sys.inputname={'io';'vg';'d'};
  sys.outputname={'vo'};

  vo_io=tf(sys(1,1)); %Output impedance transfer
                      %function vo(s)/io(s)
  vo_vg=tf(sys(1,2)); %vo(s)/vg(s)
  vo_d=tf(sys(1,3));  %Control-to-output(vo(s)/d(s))

  %drawing the Bode diagrams
  figure(1)
  bode(vo_io),grid minor,title('vo(s)/io(s)')
```

```
figure(2)
bode(vo_vg),grid minor,title('vo(s)/vg(s)')

figure(3)
bode(vo_d),grid minor,title('vo(s)/d(s)')
```

The following transfer functions are obtained after running the program[1]:

$$\frac{\tilde{v}_o(s)}{\tilde{i}_o(s)} = -.093519\frac{\left(s + 4.785 \times 10^4\right)\left(s + 1163\right)(s^2 + 1396s + 6.882 \times 10^7)}{(s^2 + 2239s + 4.76\times10^7)(s^2 + 2767s + 1.026\times10^8)} \tag{1.11}$$

$$\frac{\tilde{v}_o(s)}{\tilde{v}_g(s)} = 391.08\frac{\left(s + 4.785 \times 10^4\right)\left(s^2 + 1473s + 7.7 \times 10^7\right)}{(s^2 + 2239s + 4.76\times10^7)(s^2 + 2767s + 1.026\times10^8)} \tag{1.12}$$

$$\frac{\tilde{v}_o(s)}{\tilde{d}(s)} = 43775\frac{\left(s + 4.785 \times 10^4\right)\left(s^2 + 1371s + 7.696 \times 10^7\right)}{(s^2 + 2239s + 4.76\times10^7)(s^2 + 2767s + 1.026\times10^8)}. \tag{1.13}$$

The Bode diagram of these transfer functions are shown in Figs. 1.22, 1.23, and 1.24.

---

[1]Tilde (~) shows the small signal values. For instance: $v_o = V_o + \tilde{v}_o$ or $i_o = I_o + \tilde{i}_o$. Capital letters (for instance $V_o$ or $I_o$ in $v_o = V_o + \tilde{v}_o$ or $i_o = I_o + \tilde{i}_o$) shows the large signal values.

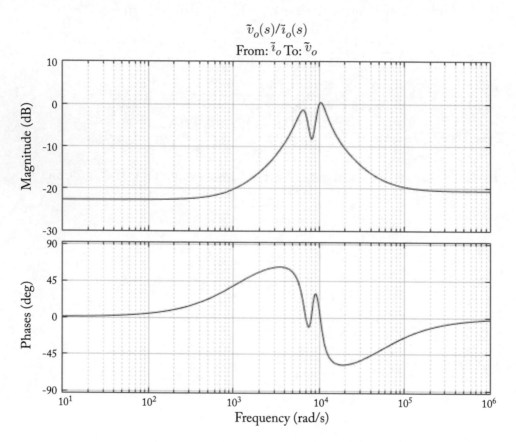

Figure 1.22: Bode diagram of $\frac{\tilde{v}_o(s)}{\tilde{i}_o(s)} = .093519\frac{(s+4.785\times10^4)(s+1163)(s^2+1396s+6.882\times10^7)}{(s^2+2239s+4.76\times10^7)(s^2+2767s+1.026\times10^8)}$.

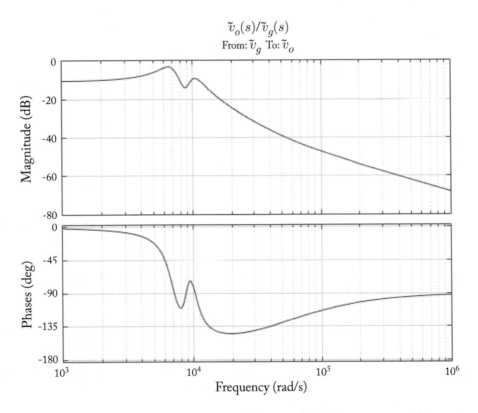

Figure 1.23: Bode diagram of $\frac{\tilde{v}_o(s)}{\tilde{v}_g(s)} = 391.08\frac{(s+4.785\times10^4)(s^2+1473s+7.7\times10^7)}{(s^2+2239s+4.76\times10^7)(s^2+2767s+1.026\times10^8)}$.

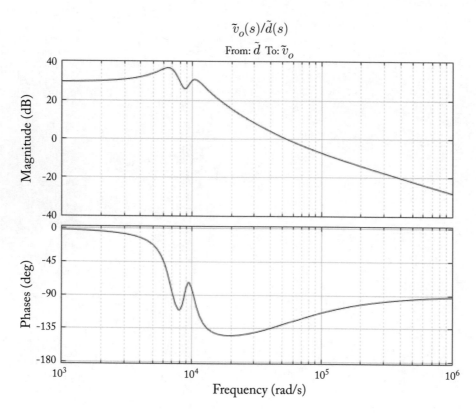

Figure 1.24: Bode diagram of $\frac{\tilde{v}_o(s)}{\tilde{d}(s)} = 43775\frac{(s+4.785\times10^4)(s^2+1371s+7.696\times10^7)}{(s^2+2239s+4.76\times10^7)(s^2+2767s+1.026\times10^8)}$.

The following block diagram can be drawn for the converter (Fig. 1.25).

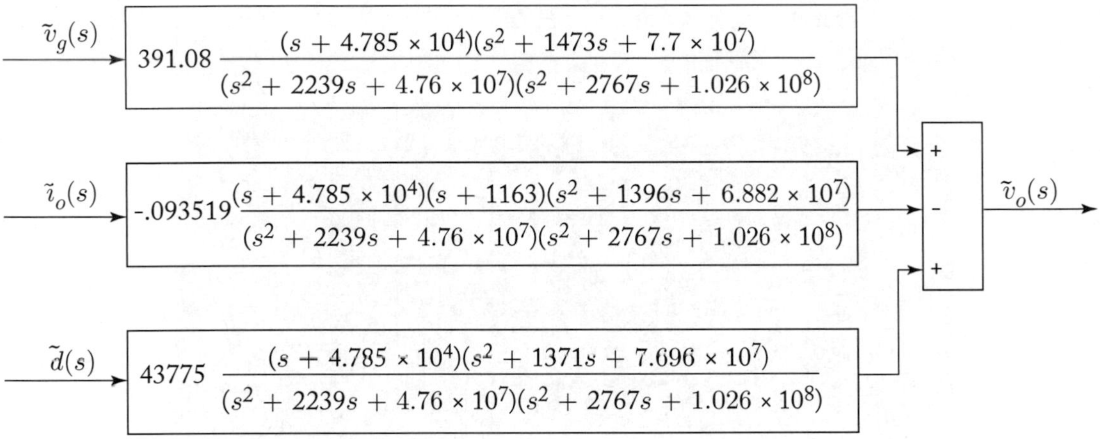

$\tilde{v}_g(s)$
$391.08 \dfrac{(s + 4.785 \times 10^4)(s^2 + 1473s + 7.7 \times 10^7)}{(s^2 + 2239s + 4.76 \times 10^7)(s^2 + 2767s + 1.026 \times 10^8)}$
$\tilde{i}_o(s)$
$-.093519 \dfrac{(s + 4.785 \times 10^4)(s + 1163)(s^2 + 1396s + 6.882 \times 10^7)}{(s^2 + 2239s + 4.76 \times 10^7)(s^2 + 2767s + 1.026 \times 10^8)}$
$\tilde{d}(s)$
$43775 \dfrac{(s + 4.785 \times 10^4)(s^2 + 1371s + 7.696 \times 10^7)}{(s^2 + 2239s + 4.76 \times 10^7)(s^2 + 2767s + 1.026 \times 10^8)}$
+
-
+
$\tilde{v}_o(s)$

Figure 1.25: Block diagram of the studied converter.

Since the injected test current ($io$ in Fig. 1.19) does not enter the positive end of output voltage, the obtained transfer function for output impedance (Equation (1.11)) must be multiplied by $-1$ to be converted to the correct form of output impedance.

## 1.4.2 VERIFICATION OF OBTAINED RESULTS

PLECS® can be used to verify the obtained results. The schematic shown in Fig. 1.26 extracts the output impedance. Extracted output impedance is shown in Fig. 1.27. The obtained result is the same as Fig. 1.22.

ZetaConverterOutputImpedance
File Edit View Simulation Format Window Help
FET2
rC1
C1
rL2
L2
.23
Duty
PWM1
Vg
rL1
L1
D
C2
rC2
R
Vo
Response
I
Perturbation

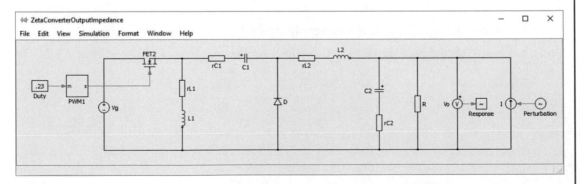

Figure 1.26: Simulation diagram to extract the output impedance $\left(\frac{\tilde{v}_o(s)}{\tilde{i}_o(s)}\right)$.

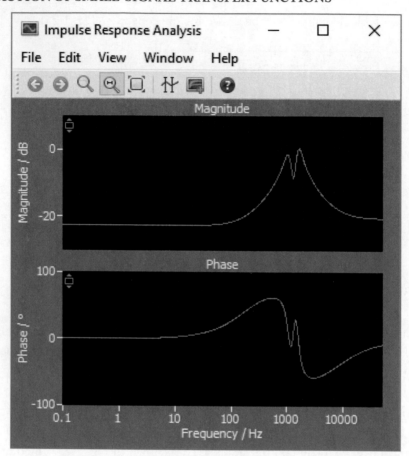

Figure 1.27: Output impedance of the studied Zeta converter (0.1–50 KHz range).

Schematics to extract the audio susceptibility $\left(\frac{\tilde{v}_o(s)}{\tilde{v}_g(s)}\right)$ and control-to-output $\left(\frac{\tilde{v}_o(s)}{\tilde{d}(s)}\right)$ are shown in Figs. 1.28 and 1.29, respectively. Analysis results are shown in Figs. 1.30 and 1.31.

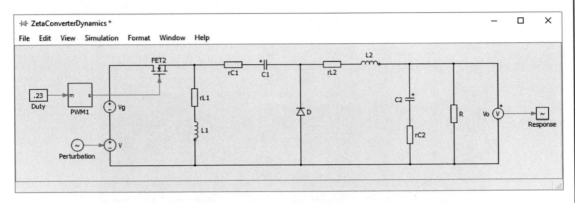

Figure 1.28: Simulation diagram to extract the audio susceptibility $\left(\frac{\tilde{v}_o(s)}{\tilde{v}_g(s)}\right)$.

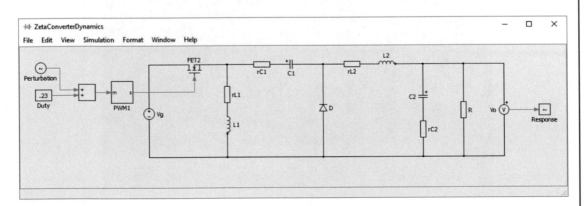

Figure 1.29: Simulation diagram to extract the control-to-output $\left(\frac{\tilde{v}_o(s)}{\tilde{d}(s)}\right)$.

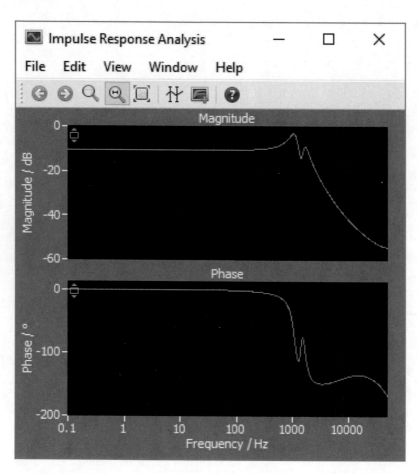

Figure 1.30: Bode diagram of audio susceptibility $\left(\frac{\tilde{v}_o(s)}{\tilde{v}_g(s)}\right)$ transfer function for studied Zeta converter.

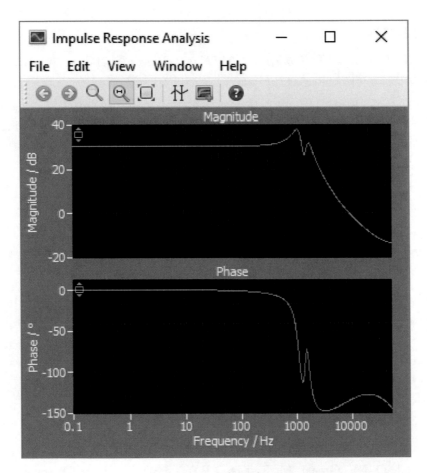

Figure 1.31: Bode diagram of control-to-output $\left(\frac{\tilde{v}_o(s)}{\tilde{d}(s)}\right)$ transfer function for studied Zeta converter.

## 1.5 CONCLUSION

This chapter studied the different techniques to extract the DC-DC converter dynamics. Both ready-to-use software packages (PLECS®) and MATLAB® programming were used to extract the converter dynamics. The next chapter focus on the input/output impedance of DC-DC converters.

# REFERENCES

Ericson, R. and Maksimovic D. *Fundamentals of Power Electronics*, 2nd ed., pp. 161–165, Norwell, Kluwer Academic Publisher, 2001. DOI: 10.1007/b100747. 1

Akbarabadi, S. A., Atighechi, H., and Jatskevich, J. Circuit-averaged and state-space-averaged-value modeling of second order flyback converter in CCM and DCM including conduction losses. *4th International Conference on Power Engineering, Energy and Electrical Drives*, pp. 995–1000, Istanbul, 2013. DOI: 10.1109/powereng.2013.6635746. 1

Asadi, F. and Eguchi, K. *Dynamics and Control of DC-DC Converters*, pp. 89–145, Morgan & Claypool, San Rafael, 2018. DOI: 10.2200/s00828ed1v01y201802pel010. 1, 16

Suntio, T. *Dynamic Profile of Switched Mode Converter: Modeling, Analysis and Control*, pp. 17–37, John Wiley & Sons, NJ, 2009. DOI: 10.1002/9783527626014. 1

Middlebrook, R. D. and Cuk, S. A general unified approach to modeling switching-converter power stages. *International Journal of Electronics Theoretical and Experimental*, pp. 521–550, 1977. DOI: 10.1109/pesc.1976.7072895. 1

Tymerski, R. and Vorperian, V. Generation, classification and analysis of switched-mode DC-DC converters by the use of converter cells. *Telecommunications Energy Conference*, pp. 181–195, 1986. DOI: 10.1109/intlec.1986.4794425. 1

Chen, J. and Ngo, K. D. T. Alternate forms of the PWM switch model in discontinuous conduction mode. *IEEE Transactions on Aerospace Electronic Systems*, pp. 754–758, 2001. DOI: 10.1109/7.937489. 1

Sun, J., Mitchell, D. M., Greuel, M. F., Krein, P. T., and Bass, R. M. Average modeling of PWM converters in discontinuous modes. *IEEE Transactions on Power Electronics*, pp. 482–492, 2001. 1

Cuk, S. and Middlebrook, R. D. A general unified approach in modeling switching DC-to-DC converters in discontinuous conduction mode. *Proc. IEEE Power Electronics Special Conference*, pp. 36–57, 1977. DOI: 10.1109/pesc.1977.7070802. 1

Suntio, T. Unified average and small-signal modeling of direct on-time control. *IEEE Transactions on Industrial Electronics*, pp. 287–295, 2006. DOI: 10.1109/tie.2005.862221. 2

Maksimovic, D., Stankovic, A. M., Tottuvelil, V. J., and Verghese, G. C. Modeling and simulation of power electronic converters. *Proc. IEEE*, vol. 89, no. 6, pp. 898–912, 2001. DOI: 10.1109/5.931486. 2

Allmeling, J. H. and Hammer, W. P. PLECS—piece-wise linear electrical circuit simulation for Simulink. *Proc. of the IEEE International Conference on Power Electronics and Drive Systems*, pp. 355–360, 1999. DOI: 10.1109/peds.1999.794588. 2

CHAPTER 2

# On the Extraction of Input and Output Impedance of PWM DC-DC Converters

## 2.1 INTRODUCTION

The input impedance of a DC-DC converter is the impedance seen from the input DC source. The output impedance is defined as the output voltage response of converter (see Fig. 2.1) for the excitation of current $i_Z$ at constant input voltage $v_G$ and duty ratio $D$. In some descriptions, the output impedance includes the load conductance $G$, in others it does not.

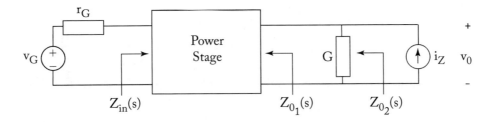

Figure 2.1: Two variants of the output impedance converter.

The input impedance of converter helps the designer to select the suitable input DC source. The input impedance of the converter must be much larger than the output impedance of the input DC source. In this case the input DC source voltage reaches the converter with low dissipations.

The output impedance of the converter is even more important than the input impedance. Output impedance must be as low as possible. Output impedance of the converter is especially important if the converter supplies a low-voltage, high-current load, with large values of output current slew rate. The most representative example of such a load is a processor in modern computer systems. The processor requires about 1.0 V (or even less) and drawn current is typically over 100 A. Current slew rates may approach 300 $\frac{A}{\mu s}$ [VRM, 2009, Singh and Khambadkone, 2008]. According to the given numbers, the processor can be modeled as a 10 mΩ resistor (or lower). The output resistance of converter should be substantially lower than 10 mΩ to ensure a good efficiency. Usually a buck converter is used to supply the processor. The output impedance of the buck converter supplying the processor (or other type of DC-DC converters) can be reduced with the aid of negative feedback. The relation between the open-loop output impedance ($Z_{O,OL}$) and closed-loop output impedance ($Z_{O,CL}$) is:

$$Z_{O,CL} = \frac{Z_{O,OL}}{1 + K_L}, \tag{2.1}$$

where $K_L$ is the loop gain [Erickson and Maksimovic, 2002, Yao et al., 2003, Ahmadi et al., 2010].

The buck, buck-boost, and boost converter are the most popular types of converters. Their input/output characteristics are the object of considerations in this chapter. Well-known references such as Erickson and Maksimovic [2002] calculate the output impedance of converter for the ideal case only, i.e., a converter without any parasitic resistance of inductors, capacitors, and switches. This chapter extracts the input/output impedance in presence of parasitic resistances.

The chapter is organized as follows. The open-loop input/output impedance of buck, buck-boost, and boost converters are derived with the aid of averaging and linearization in the second, third, and fourth sections, respectively. Input/output impedance of other types of DC-DC converters can be extracted in the same way.

## 2.2   BUCK CONVERTER

Schematic of the PWM buck converter is shown in Fig. 2.2. The working priciples of the buck converter can be found in standard text books such as Erickson and Maksimovic [2002], [2008]. $rg$, $rL$, and $rC$ show the internal resistance of the input DC source, inductor ESR, and capacitor ESR, respectively.

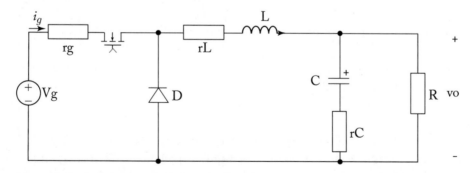

Figure 2.2: Schematic of PWM buck converter.

When the MOSFET switch is closed, the diode is reverse-biased and the equivalent circuit of Fig. 2.3 applies. $rds$ shows the MOSFET drain-source resistance. $io$ is a fictitious current source added to the circuit in order to measure the output impedance $\left(Z_o(s) = \frac{vo(s)}{io(s)}\right)$.

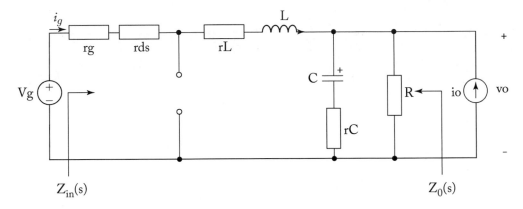

Figure 2.3: Equivalent circuit of buck converter with closed MOSFET.

According to Fig. 2.3, the circuit differential equations can be written as:

$$\frac{di_L(t)}{dt} = \frac{1}{L}\left(-\left(r_g + r_{ds} + r_L + \frac{R \times r_C}{R + r_C}\right)i_L - \frac{R}{R + r_C}v_C - \frac{R \times r_C}{R + r_C}i_O + v_g\right) \quad (2.2)$$

$$\frac{dv_C(t)}{dt} = \frac{1}{C}\left(\frac{R}{R + r_C}i_L - \frac{1}{R + r_C}v_C + \frac{R}{R + r_C}i_O\right) \quad (2.3)$$

$$i_g = i_L \quad (2.4)$$

$$v_o = r_C C\frac{dv_C}{dt} + v_C = \frac{R \times r_C}{R + r_C}i_L + \frac{R}{R + r_C}v_C + \frac{R \times r_C}{R + r_C}i_O. \quad (2.5)$$

When the MOSFET is open, the diode becomes forward-biased to carry the inductor current and the equivalent circuit of Fig. 2.4 applies. $rD$ and $VD$ show the diode resistance and diode forward voltage drop, respectively.

According to Fig. 2.4, the circuit differential equations can be written as:

$$\frac{di_L(t)}{dt} = \frac{1}{L}\left(-\left(r_D + r_L + \frac{R \times r_C}{R + r_C}\right)i_L - \frac{R}{R + r_C}v_C - \frac{R \times r_C}{R + r_C}i_O - v_D\right) \quad (2.6)$$

$$\frac{dv_C(t)}{dt} = \frac{1}{C}\left(\frac{R}{R + r_C}i_L - \frac{1}{R + r_C}v_C + \frac{R}{R + r_C}i_O\right) \quad (2.7)$$

$$i_g = 0 \quad (2.8)$$

$$v_o = r_C C\frac{dv_C}{dt} + v_C = \frac{R \times r_C}{R + r_C}i_L + \frac{R}{R + r_C}v_C + \frac{R \times r_C}{R + r_C}i_O. \quad (2.9)$$

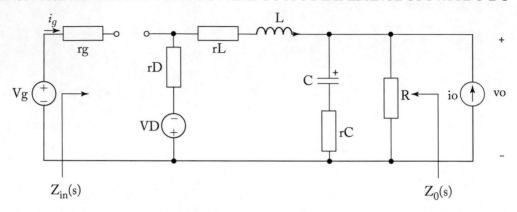

Figure 2.4: Equivalent circuit of buck converter with open MOSFET.

SSA is one of the most important tools to study the dynamics of converters operating in CCM. SSA is used in this chapter to extract the input/output impedance of studied converters. As said in the previous chapter, SSA has two important steps: averaging and linearization. The SSA procedure can be summarized as follows.

**Step 1** Circuit differential equations are written for different working modes (i.e., on/off state of semiconductor switches).

**Step 2** Equations are time averaged over one period.

**Step 3** Steady-state operating points are calculated by equating the derivative terms to zero.

**Step 4** The averaged equations are linearized around the steady-state operating point found in the Step 3.

Applying SSA to Equations (2.2)–(2.9) leads to six different transfer functions: $\frac{i_g(s)}{d(s)}$, $\frac{i_g(s)}{v_g(s)}$, $\frac{i_g(s)}{i_o(s)}$, $\frac{v_o(s)}{d(s)}$, $\frac{v_o(s)}{v_g(s)}$, and $\frac{v_o(s)}{i_o(s)}$. Open-loop input and output impedance of the converter is extracted with the aid of $\frac{1}{\frac{i_g(s)}{v_g(s)}}$ and $\frac{v_o(s)}{i_o(s)}$, respectively.

MATLAB® can be very helpful to do the mathematical machinery of SSA. The following program extracts the small-signal transfer functions of a buck converter with component values, as shown in Table 2.1.

Table 2.1: The buck converter parameters (see Fig. 2.2)

| | Nominal Value |
|---|---|
| Output voltage, Vo | 20 V |
| Duty ratio, D | 0.4 |
| Input DC source voltage, Vg | 50 V |
| Input DC source internal resistance, rg | 0.01 $\Omega$ |
| MOSFET drain-source resistance, rds | 40 m$\Omega$ |
| Capacitor, C | 100 $\mu$F |
| Capacitor Equivaluent Series Resistance (ESR), rC | 0.05 $\Omega$ |
| Inductor, L | 400 $\mu$H |
| Inductor ESR, rL | 50 m$\Omega$ |
| Diode voltage drop, vD | 0.7 V |
| Diode forward resistance, rD | 10 m$\Omega$ |
| Load resistor, R | 20 $\Omega$ |
| Switching Frequency, Fsw | 20 KHz |

```
%This program calculates the input and
%output impedance of the Buck converter.

clc

clear all
syms vg rg d rL L rC C R vC iL rds rD vD io

%Converter Dynamical equations
%M1: diL/dt for closed MOSFET.
%M2: dvC/dt for closed MOSFET.
%M3: current of input DC source for closed MOSFET.
%M4: output voltage of converter for closed MOSFET.

%M5: diL/dt for open MOSFET.
%M6: dvC/dt for open MOSFET.
%M7: current of input DC source for open MOSFET.
%M8: output voltage of converter for open MOSFET.
```

```
M1=(-(rg+rds+rL+R*rC/(R+rC))*iL-R/(R+rC)*vC-R*rC/(R+rC)*io+vg)/L;
M2=(R/(R+rC)*iL-1/(R+rC)*vC+R/(R+rC)*io)/C;
M3=iL;
M4=R*rC/(R+rC)*iL+R/(R+rC)*vC+R*rC/(R+rC)*io;

M5=(-(rD+rL+R*rC/(R+rC))*iL-R/(R+rC)*vC-R*rC/(R+rC)*io-vD)/L;
M6=(R/(R+rC)*iL-1/(R+rC)*vC+R/(R+rC)*io)/C;
M7=0;
M8=R*rC/(R+rC)*iL+R/(R+rC)*vC+R*rC/(R+rC)*io;

%Averaged Equations
diL_dt_ave=simplify(M1*d+M5*(1-d));
dvC_dt_ave=simplify(M2*d+M6*(1-d));
ig_ave=simplify(M3*d+M7*(1-d));
vo_ave=simplify(M4*d+M8*(1-d));

%DC Operating Point
DC=solve(diL_dt_ave==0,dvC_dt_ave==0,'iL','vC');
IL=DC.iL;
VC=DC.vC;

%Linearization
A11=simplify(subs(diff(diL_dt_ave,iL),[iL vC io],[IL VC 0]));
A12=simplify(subs(diff(diL_dt_ave,vC),[iL vC io],[IL VC 0]));
A21=simplify(subs(diff(dvC_dt_ave,iL),[iL vC io],[IL VC 0]));
A22=simplify(subs(diff(dvC_dt_ave,vC),[iL vC io],[IL VC 0]));
AA=[A11 A12;A21 A22];

B11=simplify(subs(diff(diL_dt_ave,io),[iL vC io],[IL VC 0]));
B12=simplify(subs(diff(diL_dt_ave,vg),[iL vC io],[IL VC 0]));
B13=simplify(subs(diff(diL_dt_ave,d),[iL vC io],[IL VC 0]));

B21=simplify(subs(diff(dvC_dt_ave,io),[iL vC io],[IL VC 0]));
B22=simplify(subs(diff(dvC_dt_ave,vg),[iL vC io],[IL VC 0]));
B23=simplify(subs(diff(dvC_dt_ave,d),[iL vC io],[IL VC 0]));

BB=[B11 B12 B13;B21 B22 B23];

C11=simplify(subs(diff(ig_ave,iL),[iL vC io],[IL VC 0]));
```

```
C12=simplify(subs(diff(ig_ave,vC),[iL vC io],[IL VC 0]));

C21=simplify(subs(diff(vo_ave,iL),[iL vC io],[IL VC 0]));
C22=simplify(subs(diff(vo_ave,vC),[iL vC io],[IL VC 0]));
CC=[C11 C12; C21 C22];

D11=simplify(subs(diff(ig_ave,io),[iL vC io],[IL VC 0 ]));
D12=simplify(subs(diff(ig_ave,vg),[iL vC io],[IL VC 0]));
D13=simplify(subs(diff(ig_ave,d),[iL vC io],[IL VC 0]));

D21=simplify(subs(diff(vo_ave,io),[iL vC io],[IL VC 0 ]));
D22=simplify(subs(diff(vo_ave,vg),[iL vC io],[IL VC 0]));
D23=simplify(subs(diff(vo_ave,d),[iL vC io],[IL VC 0]));
DD=[D11 D12 D13;D21 D22 D23];

%Components Values
%Variables have underline are used to
%store the numeric values of components
%Variables without underline are symbolic variables.
%for example:
%L: symbolic vvariable shows the inductor inductance
%L_: numeric variable shows the inductor inductance value.
L_=400e-6;
rL_=.05;
C_=100e-6;
rC_=.05;
rds_=.04;
rD_=.01;
VD_=.7;
D_=.4;
VG_=50;
rg_=.01;
R_=20;

AA_=eval(subs(AA,[vg rg rds rD vD rL L rC C R d io],
    [VG_ rg_rds_ rD_ VD_ rL_ L_ rC_ C_ R_ D_ 0]));
BB_=eval(subs(BB,[vg rg rds rD vD rL L rC C R d io],
    [VG_ rg_rds_ rD_ VD_ rL_ L_ rC_ C_ R_ D_ 0]));
CC_=eval(subs(CC,[vg rg rds rD vD rL L rC C R d io],
```

```
    [VG_ rg_rds_ rD_ VD_ rL_ L_ rC_ C_ R_ D_ 0]));
DD_=eval(subs(DD,[vg rg rds rD vD rL L rC C R d io],
    [VG_ rg_rds_ rD_ VD_ rL_ L_ rC_ C_ R_ D_ 0]));

sys=ss(AA_,BB_,CC_,DD_);
sys.stateName={'iL','vC'};
sys.inputname={'io','vg','d'};
sys.outputname={'ig','vo'};

ig_io=sys(1,1);
ig_vg=sys(1,2);
ig_d=sys(1,3);

vo_io=sys(2,1);
vo_vg=sys(2,2);
vo_d=sys(2,3);

Zin=1/ig_vg; %input impedance
Zout=vo_io;  %output impedance

%Draws the bode diagram of input/output impedance
figure(1)
bode(Zin), grid minor

figure(2)
bode(Zout), grid minor

%Display the DC operating point of converter
disp('steady state operating point of converter')
disp('IL')
disp(eval(subs(IL,[vg rg rds rD vD rL L rC C R d io],
    [VG_ rg_rds_ rD_ VD_ rL_ L_ rC_ C_ R_ D_ 0])));
disp('VC')
disp(eval(subs(VC,[vg rg rds rD vD rL L rC C R d io],
    [VG_ rg_rds_ rD_ VD_ rL_ L_ rC_ C_ R_ D_ 0])));
```

The program gives the following results (OL sub-script stands for Open Loop):

$$\frac{v_o(s)}{d(s)} = 6316.8 \frac{s + 2 \times 10^5}{s^2 + 813.4s + 2.503 \times 10^7} \tag{2.10}$$

$$Z_{in,OL}(s) = \frac{v_g(s)}{i_g(s)} = 0.0025 \frac{s^2 + 813.4s + 2.503 \times 10^7}{s + 498.8} \tag{2.11}$$

$$Z_{o,OL}(s) = \frac{v_o(s)}{i_o(s)} = 0.049875 \frac{(s + 2 \times 10^5)(s + 190)}{s^2 + 813.4s + 2.503 \times 10^7}. \tag{2.12}$$

Bode diagram of control-to-output transfer function, open-loop input impedance, and open-loop output impedance is shown in Figs. 2.5, 2.6, and 2.7, respectively.

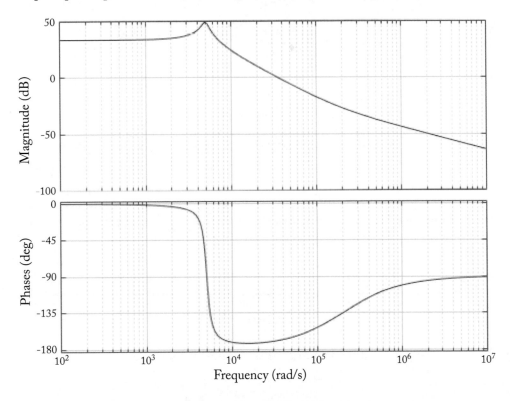

Figure 2.5: Control-to-output transfer function of the studied buck converter.

The block diagram shown in Fig. 2.8 can be drawn for the studied buck converter. We want to study the effect of feedback on output impedance.

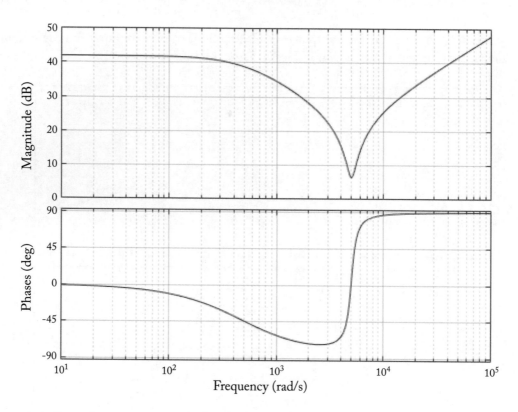

Figure 2.6: Open-loop input impedance of the studied buck converter.

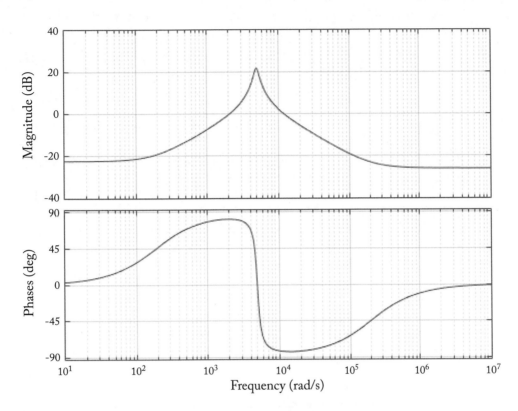

Figure 2.7: Open-loop output impedance of the studied buck converter.

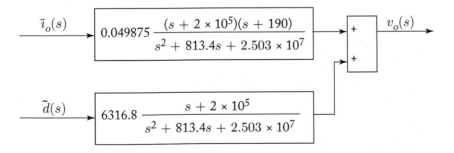

Figure 2.8: Dynamical model of the studied buck converter. Input voltage variations ($v_g$) are ignored.

Consider a simple feedback loop shown in Fig. 2.9. Assume that the controller is a simple I-type controller $\left(C\left(s\right) = \frac{4.85}{s}\right)$. Figure 2.10 shows the step response of the closed loop.

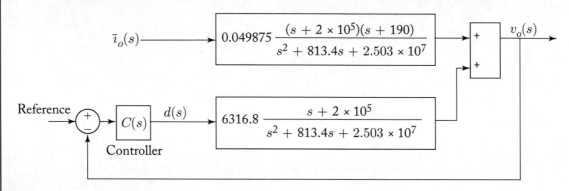

Figure 2.9: Voltage Mode (VM) control of the studied buck converter.

According to Fig. 2.9, the closed-loop output impedance $(Z_{o,CL}\left(s\right))$ is:

$$Z_{o,CL}\left(s\right) = Z_{o,OL}\left(s\right) \times \frac{1}{1 + C(s) \times \frac{vo(s)}{d(s)}} \qquad (2.13)$$

$$Z_{o,CL}\left(s\right) = 0.049875 \times \frac{\left(s + 2 \times 10^5\right)\left(s + 190\right)}{s^2 + 813.4s + 2.503 \times 10^7}$$

$$\times \frac{1}{1 + \frac{4.85}{s} \times 6316.8 \times \frac{\left(s + 2 \times 10^5\right)}{s^2 + 813.4s + 2.503 \times 10^7}}$$

$$= \frac{0.04988s^5 + 10^4 s^4 + 1.13 \times 10^7 s^3 + 2.515 \times 10^{11} s^2 + 4.744 \times 10^{13} s}{s^5 + 1627s^4 + 5.075 \times 10^7 s^3 + 4.687 \times 10^{10} + 6.323 \times 10^{14} s + 1.534 \times 10^{17}}.$$

$$(2.14)$$

Figure 2.11 is a comparison between the open-loop output impedance $(Z_{o,OL}(s)$, Equation (2.12)) and the closed-loop output impedance $(Z_{o,CL}(s)$, Equation (2.14)). The closed-loop output impedance is reduced at low-frequency portion of the graph. Reduction of output impedance is one of the desired properties of feedback control.

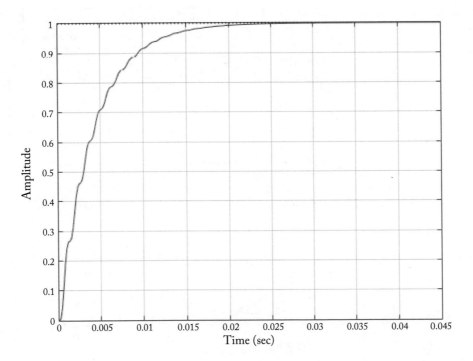

Figure 2.10: Step response of closed-loop control system shown in Fig. 2.9 with $C\left(s\right) = \frac{4.85}{s}$.

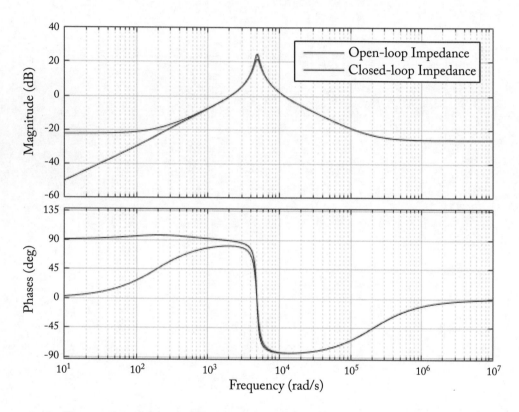

Figure 2.11: Comparison of open-loop output impedance with closed-loop output impedance for the studied buck converter.

There are many efforts presented in the literature to achieve satisfactory output impedance of PWM DC-DC converters, especially buck type. The methods can be categorized into two groups:

- sophisticated design of control loops in the converter [Chen et al., 2007, Lee et al., 2008, Xiao et al., 2008, Qahouq and Arikatla, 2011] and

- modifications of the basic structure of the power stage [Singh and Khambadkone, 2008].

The starting point of the first method is the precise description of the converter, in particular the use of accurate formulas for open-loop output impedance. The given program can do this step.

## 2.3   BUCK-BOOST CONVERTER

Schematic of the buck-boost converter is shown in Fig. 2.12.

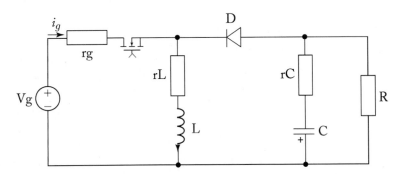

Figure 2.12: Schematic of PWM buck-boost converter.

When the MOSFET is closed, the diode is reverse biased and Fig. 2.13 is an equivalent circuit. According to Fig. 2.13, the circuit differential equations can be written as:

$$\frac{di_L(t)}{dt} = \frac{1}{L}\left(-\left(r_g + r_{ds} + r_L\right)i_L + v_g\right) \tag{2.15}$$

$$\frac{dv_C(t)}{dt} = \frac{1}{C}\left(\frac{R}{R + r_C}i_O - \frac{1}{R + r_C}v_C\right) \tag{2.16}$$

$$i_g = i_L \tag{2.17}$$

$$v_o = \frac{R}{R + r_C}v_C + \frac{R \times r_C}{R + r_C}i_O. \tag{2.18}$$

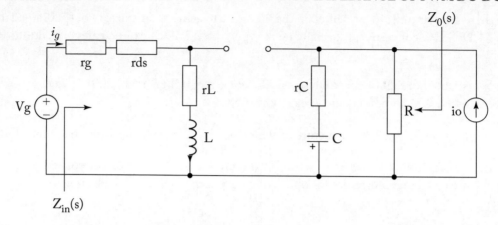

Figure 2.13: Equivalent circuit of buck-boost converter with closed MOSFET.

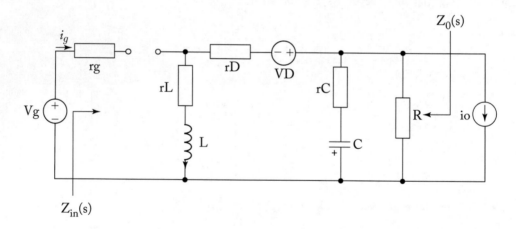

Figure 2.14: Equivalent circuit of buck-boost converter with open MOSFET.

When the MOSFET switch is opened, the diode becomes forward-biased. Figure 2.14 shows the equivalent circuit for this case. According to Fig. 2.14, the circuit differential equations can be written as:

$$\frac{di_L(t)}{dt} = \frac{1}{L}\left(-\left(r_D + r_L + \frac{R \times r_C}{R + r_C}\right)i_L - \frac{R}{R + r_C}v_C - \frac{R \times r_C}{R + r_C}i_O - v_D\right) \quad (2.19)$$

$$\frac{dv_C(t)}{dt} = \frac{1}{C}\left(\frac{R}{R + r_C}i_L - \frac{1}{R + r_C}v_C + \frac{R}{R + r_C}i_O\right) \quad (2.20)$$

$$i_g = 0 \qquad (2.21)$$

$$v_o = \frac{R \times r_C}{R + r_C} i_L + \frac{R}{R + r_C} v_C + \frac{R \times r_C}{R + r_C} i_O + V_D. \qquad (2.22)$$

Consider a buck-boost converter with component values as shown in Table 2.2. The following program extracts the small signal transfer function of the assumed converter.

Table 2.2: The buck-boost converter parameters (see Fig. 2.12)

|  | Nominal Value |
| --- | --- |
| Output voltage, Vo | -16 V |
| Duty ratio, D | 0.4 |
| Input DC source voltage, Vg | 24 V |
| Input DC source internal resistance, rg | 0.1 Ω |
| MOSFET drain-source resistance, rds | 40 mΩ |
| Capacitor, C | 80 μF |
| Capacitor Equivaluent Series Resistance (ESR), rC | 0.05 Ω |
| Inductor, L | 20 μH |
| Inductor ESR, rL | 10 mΩ |
| Diode voltage drop, vD | 0.7 V |
| Diode forward resistance, rD | 10 mΩ |
| Load resistor, R | 5 Ω |
| Switching Frequency, Fsw | 100 KHz |

```
%This program calculates the input and output
%impedance of the Buck-Boost converter.

clc

clear all
syms vg rg d rL L rC C R vC iL rds rD vD io

%Converter Dynamical equations
%M1: diL/dt for closed MOSFET.
%M2: dvC/dt for closed MOSFET.
```

```
%M3: current of input DC source for closed MOSFET.
%M4: output voltage of converter for closed MOSFET.

%M5: diL/dt for open MOSFET.
%M6: dvC/dt for open MOSFET.
%M7: current of input DC source for open MOSFET.
%M8: output voltage of converter for open MOSFET.

M1=(-(rg+rds+rL)*iL+vg)/L;
M2=(R/(R+rC)*io-vC/(R+rC))/C;
M3=iL;
M4=R*rC/(R+rC)*io+R/(R+rC)*vC;

M5=(-(rL+rD+rC*R/(R+rC))*iL-R/(R+rC)*vC-R*rC/(R+rC)*io-vD)/L;
M6=(R/(R+rC)*iL-1/(R+rC)*vC+R/(R+rC)*io)/C;
M7=0;
M8=rC*R/(rC+R)*iL+R/(R+rC)*vC+R*rC/(R+rC)*io+vD;

%Averaged Equations
diL_dt_ave=simplify(M1*d+M5*(1-d));
dvC_dt_ave=simplify(M2*d+M6*(1-d));
ig_ave=simplify(M3*d+M7*(1-d));
vo_ave=simplify(M4*d+M8*(1-d));

%DC Operating Point
DC=solve(diL_dt_ave==0,dvC_dt_ave==0,'iL','vC');
IL=DC.iL;
VC=DC.vC;

%Linearization
A11=simplify(subs(diff(diL_dt_ave,iL),[iL vC io],[IL VC 0]));
A12=simplify(subs(diff(diL_dt_ave,vC),[iL vC io],[IL VC 0]));
A21=simplify(subs(diff(dvC_dt_ave,iL),[iL vC io],[IL VC 0]));
A22=simplify(subs(diff(dvC_dt_ave,vC),[iL vC io],[IL VC 0]));
AA=[A11 A12;A21 A22];

B11=simplify(subs(diff(diL_dt_ave,io),[iL vC io],[IL VC 0]));
B12=simplify(subs(diff(diL_dt_ave,vg),[iL vC io],[IL VC 0]));
B13=simplify(subs(diff(diL_dt_ave,d),[iL vC io],[IL VC 0]));
```

```
B21=simplify(subs(diff(dvC_dt_ave,io),[iL vC io],[IL VC 0]));
B22=simplify(subs(diff(dvC_dt_ave,vg),[iL vC io],[IL VC 0]));
B23=simplify(subs(diff(dvC_dt_ave,d),[iL vC io],[IL VC 0]));

BB=[B11 B12 B13;B21 B22 B23];

C11=simplify(subs(diff(ig_ave,iL),[iL vC io],[IL VC 0]));
C12=simplify(subs(diff(ig_ave,vC),[iL vC io],[IL VC 0]));

C21=simplify(subs(diff(vo_ave,iL),[iL vC io],[IL VC 0]));
C22=simplify(subs(diff(vo_ave,vC),[iL vC io],[IL VC 0]));
CC=[C11 C12; C21 C22];

D11=simplify(subs(diff(ig_ave,io),[iL vC io],[IL VC 0 ]));
D12=simplify(subs(diff(ig_ave,vg),[iL vC io],[IL VC 0]));
D13=simplify(subs(diff(ig_ave,d),[iL vC io],[IL VC 0]));

D21=simplify(subs(diff(vo_ave,io),[iL vC io],[IL VC 0 ]));
D22=simplify(subs(diff(vo_ave,vg),[iL vC io],[IL VC 0]));
D23=simplify(subs(diff(vo_ave,d),[iL vC io],[IL VC 0]));
DD=[D11 D12 D13;D21 D22 D23];

%Components Values
%Variables have underline are used to
%store the numeric values of components
%Variables without underline are symbolic variables.
%for example:
%L: symbolic vvariable shows the inductor inductance
%L_: numeric variable shows the inductor inductance value.
L_=20e-6;
rL_=.01;
C_=80e-6;
rC_=.05;
rds_=.04;
rD_=.01;
VD_=.7;
D_=.4;
VG_=24;
```

```
rg_=.1;
R_=5;

AA_=eval(subs(AA,[vg rg rds rD vD rL L rC C R d io],
    [VG_ rg_rds_ rD_ VD_ rL_ L_ rC_ C_ R_ D_ 0]));
BB_=eval(subs(BB,[vg rg rds rD vD rL L rC C R d io],
    [VG_ rg_rds_ rD_ VD_ rL_ L_ rC_ C_ R_ D_ 0]));
CC_=eval(subs(CC,[vg rg rds rD vD rL L rC C R d io],
    [VG_ rg_rds_ rD_ VD_ rL_ L_ rC_ C_ R_ D_ 0]));
DD_=eval(subs(DD,[vg rg rds rD vD rL L rC C R d io],
    [VG_ rg_rds_ rD_ VD_ rL_ L_ rC_ C_ R_ D_ 0]));

sys=ss(AA_,BB_,CC_,DD_);
sys.stateName={'iL','vC'};
sys.inputname={'io','vg','d'};
sys.outputname={'ig','vo'};

ig_io=sys(1,1);
ig_vg=sys(1,2);
ig_d=sys(1,3);

vo_io=sys(2,1);
vo_vg=sys(2,2);
vo_d=sys(2,3);

Zin=1/ig_vg; %input impedance
Zout=vo_io;   %output impedance

%Draws the bode diagram of input/output impedance
figure(1)
bode(Zin), grid minor

figure(2)
bode(Zout), grid minor

%Display the DC operating point of converter
disp('steady state operating point of converter')
disp('IL')
disp(eval(subs(IL,[vg rg rds rD vD rL L rC C R d io],
```

```
    [VG_ rg_rds_ rD_ VD_ rL_ L_ rC_ C_ R_ D_ 0])));
disp('VC')
disp(eval(subs(VC,[vg rg rds rD vD rL L rC C R d io],
    [VG_ rg_rds_ rD_ VD_ rL_ L_ rC_ C_ R_ D_ 0])));
```

The program gives the following results:

$$\frac{v_o(s)}{d(s)} = -0.94123 \frac{\left(s + 1.267 \times 10^5\right)\left(s - 1.168 \times 10^5\right)}{s^2 + 7560s + 2.332 \times 10^8} \tag{2.23}$$

$$Z_{in}(s) - \frac{v_g(s)}{i_g(s)} = 0.000125 \frac{s^2 + 7560s + 2.332 \times 10^8}{s + 2475} \tag{2.24}$$

$$Z_o(s) = \frac{v_o(s)}{i_o(s)} = 0.049505 \frac{\left(s + 2.5 \times 10^5\right)\left(s + 4194\right)}{s^2 + 7560s + 2.332 \times 10^8}. \tag{2.25}$$

Bode diagram of control-to-output transfer function, open-loop input impedance, and open-loop output impedance is shown in Figs. 2.15, 2.16, and 2.17, respectively.

## 2.4    BOOST CONVERTER

Schematic of the boost converter is shown in Fig. 2.18. When the MOSFET is closed, the diode is reverse biased and Fig. 2.19 is an equivalent circuit.

According to Fig. 2.19, the circuit differential equations can be written as:

$$\frac{di_L(t)}{dt} = \frac{1}{L}\left(-\left(r_g + r_{ds} + r_L\right)i_L + v_g\right) \tag{2.26}$$

$$\frac{dv_C(t)}{dt} = \frac{1}{C}\left(-\frac{1}{R + r_C}v_C + \frac{R}{R + r_C}i_o\right) \tag{2.27}$$

$$i_g = i_L \tag{2.28}$$

$$v_o = \frac{R}{R + r_C}v_C + \frac{R \times r_C}{R + r_C}i_o. \tag{2.29}$$

When the MOSFET switch is opened, the diode becomes forward-biased. Figure 2.20 shows the equivalent circuit for this case.

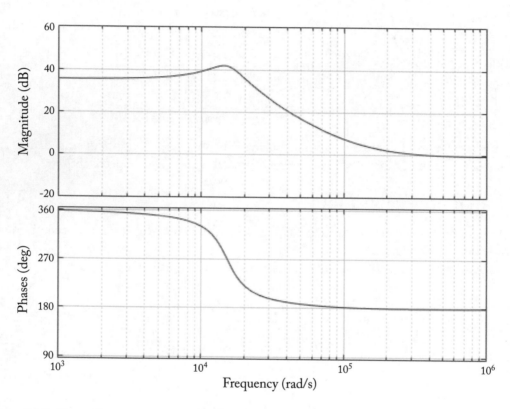

Figure 2.15: Control-to-output transfer function of the studied buck-boost converter.

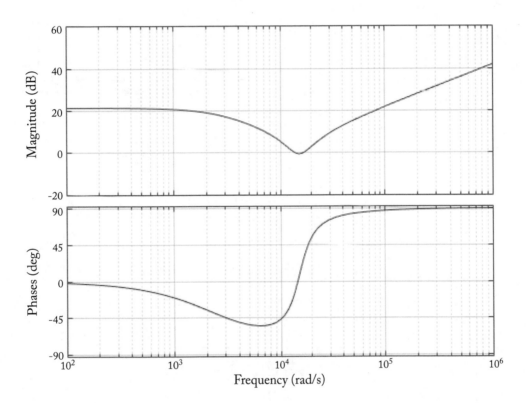

Figure 2.16: Open-loop input impedance of buck-boost converter.

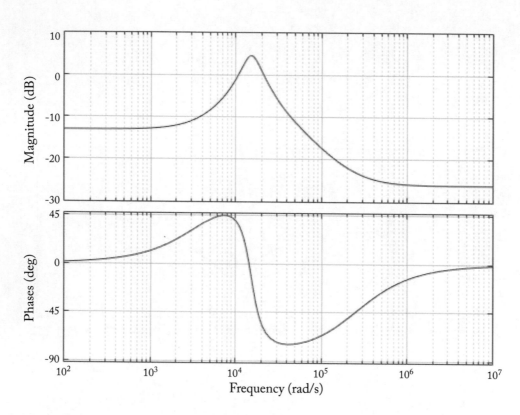

Figure 2.17: Open-loop output impedance of buck-boost converter.

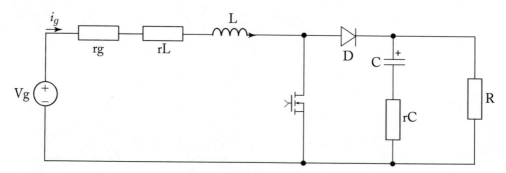

Figure 2.18: Schematic of PWM boost converter.

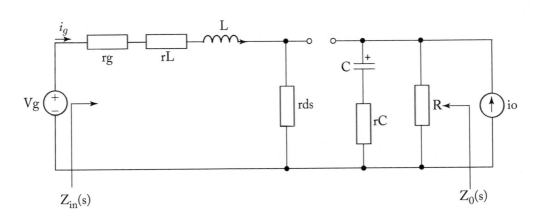

Figure 2.19: Equivalent circuit of boost converter with closed MOSFET.

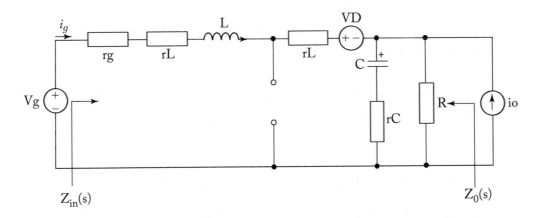

Figure 2.20: Equivalent circuit of boost converter with open MOSFET.

According to Fig. 2.20, the circuit differential equations can be written as:

$$\frac{di_L(t)}{dt} = \frac{1}{L}\left(-\left(r_g + r_L + r_D + \frac{R \times r_C}{R + r_C}\right)i_L - \frac{R}{R + r_C}v_C - \frac{R \times r_C}{R + r_C}i_O + v_g - v_D\right)$$

(2.30)

$$\frac{dv_C(t)}{dt} = \frac{1}{C}\left(\frac{R}{R + r_C}i_L - \frac{1}{R + r_C}v_C + \frac{R}{R + r_C}i_O\right)$$

(2.31)

$$i_g = i_L$$

(2.32)

$$v_o = \frac{R \times r_C}{R + r_C}i_L + \frac{R}{R + r_C}v_C + \frac{R \times r_C}{R + r_C}i_O.$$

(2.33)

Consider a boost converter with component values as shown in Table 2.3. The following program extracts the small signal transfer function of the assumed converter.

Table 2.3: The boost converter parameters (see Fig. 2.18).

|  | Nominal Value |
|---|---|
| Output voltage, Vo | 30 V |
| Duty ratio, D | 0.6 |
| Input DC source voltage, Vg | 12 V |
| Input DC source internal resistance, rg | 0.1 Ω |
| MOSFET drain-source resistance, rds | 40 mΩ |
| Capacitor, C | 100 μF |
| Capacitor Equivaluent Series Resistance (ESR), rC | 0.05 Ω |
| Inductor, L | 120 μH |
| Inductor ESR, rL | 10 mΩ |
| Diode voltage drop, vD | 0.7 V |
| Diode forward resistance, rD | 10 mΩ |
| Load resistor, R | 50 Ω |
| Switching Frequency, Fsw | 25 KHz |

```
%This program calculates the input and output
%impedance of the Boost converter.

clc

clear all
syms vg rg d rL L rC C R vC iL rds rD vD io

%Converter Dynamical equations
%M1: diL/dt for closed MOSFET.
%M2: dvC/dt for closed MOSFET.
%M3: current of input DC source for closed MOSFET.
%M4: output voltage of converter for closed MOSFET.

%M5: diL/dt for open MOSFET.
%M6: dvC/dt for open MOSFET.
%M7: current of input DC source for open MOSFET.
%M8: output voltage of converter for open MOSFET.

M1=(-(rg+rL+rds)*iL+vg)/L;
M2=(-vC/(R+rC)+R/(R+rC)*io)/C;
M3=iL;
M4=R/(R+rC)*vC+R*rC/(R+rC)*io;

M5=(-(rg+rL+rD+R*rC/(R+rC))*iL-R/(R+rC)*vC-R*rC/(R+rC)*
    io+vg-vD)/L;
M6=((R/(R+rC))*iL-vC/(R+rC)+R/(R+rC)*io)/C;
M7=iL;
M8=R*rC/(R+rC)*iL-R/(R+rC)*vC+R*rC/(R+rC)*io;

%Averaged Equations
diL_dt_ave=simplify(M1*d+M5*(1-d));
dvC_dt_ave=simplify(M2*d+M6*(1-d));
ig_ave=simplify(M3*d+M7*(1-d));
vo_ave=simplify(M4*d+M8*(1-d));

%DC Operating Point
DC=solve(diL_dt_ave==0,dvC_dt_ave==0,'iL','vC');
```

```
IL=DC.iL;
VC=DC.vC;

%Linearization
A11=simplify(subs(diff(diL_dt_ave,iL),[iL vC io],[IL VC 0]));
A12=simplify(subs(diff(diL_dt_ave,vC),[iL vC io],[IL VC 0]));
A21=simplify(subs(diff(dvC_dt_ave,iL),[iL vC io],[IL VC 0]));
A22=simplify(subs(diff(dvC_dt_ave,vC),[iL vC io],[IL VC 0]));
AA=[A11 A12;A21 A22];

B11=simplify(subs(diff(diL_dt_ave,io),[iL vC io],[IL VC 0]));
B12=simplify(subs(diff(diL_dt_ave,vg),[iL vC io],[IL VC 0]));
B13=simplify(subs(diff(diL_dt_ave,d),[iL vC io],[IL VC 0]));

B21=simplify(subs(diff(dvC_dt_ave,io),[iL vC io],[IL VC 0]));
B22=simplify(subs(diff(dvC_dt_ave,vg),[iL vC io],[IL VC 0]));
B23=simplify(subs(diff(dvC_dt_ave,d),[iL vC io],[IL VC 0]));

BB=[B11 B12 B13;B21 B22 B23];

C11=simplify(subs(diff(ig_ave,iL),[iL vC io],[IL VC 0]));
C12=simplify(subs(diff(ig_ave,vC),[iL vC io],[IL VC 0]));

C21=simplify(subs(diff(vo_ave,iL),[iL vC io],[IL VC 0]));
C22=simplify(subs(diff(vo_ave,vC),[iL vC io],[IL VC 0]));
CC=[C11 C12; C21 C22];

D11=simplify(subs(diff(ig_ave,io),[iL vC io],[IL VC 0 ]));
D12=simplify(subs(diff(ig_ave,vg),[iL vC io],[IL VC 0]));
D13=simplify(subs(diff(ig_ave,d),[iL vC io],[IL VC 0]));

D21=simplify(subs(diff(vo_ave,io),[iL vC io],[IL VC 0 ]));
D22=simplify(subs(diff(vo_ave,vg),[iL vC io],[IL VC 0]));
D23=simplify(subs(diff(vo_ave,d),[iL vC io],[IL VC 0]));
DD=[D11 D12 D13;D21 D22 D23];

%Components Values
%Variables have underline are used to store
%the numeric values of components
```

```
%Variables without underline are symbolic variables.
%for example:
%L: symbolic vvariable shows the inductor inductance
%L_: numeric variable shows the inductor inductance value.
L_=120e-6;
rL_=.01;
C_=100e-6;
rC_=.05;
rds_=.04;
rD_=.01;
VD_=.7;
D_=.6;
VG_=12;
rg_=.1;
R_=50;

AA_=eval(subs(AA,[vg rg rds rD vD rL L rC C R d io],
    [VG_ rg_rds_ rD_ VD_ rL_ L_ rC_ C_ R_ D_ 0]));
BB_=eval(subs(BB,[vg rg rds rD vD rL L rC C R d io],
    [VG_ rg_rds_ rD_ VD_ rL_ L_ rC_ C_ R_ D_ 0]));
CC_=eval(subs(CC,[vg rg rds rD vD rL L rC C R d io],
    [VG_ rg_rds_ rD_ VD_ rL_ L_ rC_ C_ R_ D_ 0]));
DD_=eval(subs(DD,[vg rg rds rD vD rL L rC C R d io],
    [VG_ rg_rds_ rD_ VD_ rL_ L_ rC_ C_ R_ D_ 0]));

sys=ss(AA_,BB_,CC_,DD_);
sys.stateName={'iL','vC'};
sys.inputname={'io','vg','d'};
sys.outputname={'ig','vo'};

ig_io=sys(1,1);
ig_vg=sys(1,2);
ig_d=sys(1,3);

vo_io=sys(2,1);
vo_vg=sys(2,2);
vo_d=sys(2,3);

Zin=1/ig_vg; %input impedance
```

```
Zout=vo_io;   %output impedance

%Draws the bode diagram of input/output impedance
figure(1)
bode(Zin), grid minor

figure(2)
bode(Zout), grid minor

%Display the DC operating point of converter
disp('steady state operating point of converter')
disp('IL')
disp(eval(subs(IL,[vg rg rds rD vD rL L rC C R d io],
    [VG_ rg_rds_ rD_ VD_ rL_ L_ rC_ C_ R_ D_ 0])));
disp('VC')
disp(eval(subs(VC,[vg rg rds rD vD rL L rC C R d io],
    [VG_ rg_rds_ rD_ VD_ rL_ L_ rC_ C_ R_ D_ 0])));
```

The program gives the following results:

$$\frac{v_o(s)}{d(s)} = -0.007199\frac{\left(s + 2 \times 10^6\right)\left(s - 6.703 \times 10^4\right)}{s^2 + 1367s + 1.356 \times 10^7} \tag{2.34}$$

$$Z_{in}(s) = \frac{v_g(s)}{i_g(s)} = 0.00012\frac{s^2 + 1367s + 1.356 \times 10^7}{s + 200} \tag{2.35}$$

$$Z_o(s) = \frac{v_o(s)}{i_o(s)} = 0.049995\frac{\left(s + 2 \times 10^6\right)(s + 1160)}{s^2 + 1367s + 1.356 \times 10^7}. \tag{2.36}$$

Bode diagram of control-to-output transfer function, open-loop input impedance and open-loop output impedance is shown in Figs. 2.21, 2.22, and 2.23, respectively.

The programs given in this chapter calculates the steady-state operating point of the converter as well. The steady-state operating point of the studied boost converter is $I_L = 1.438$ A and $V_C = 28.76$ V. $I_L$ and $V_C$ show the average inductor current and average capacitor voltage, respectively.

The average current drawn from the input DC source is the same as the average current of inductor. So, the input DC source sees the converter as a $\frac{12 \text{ V}}{1.438 \text{ A}} = 8.35\ \Omega$ load. If we substitute $s = 0$ in Equation (2.35), we obtain the 8.15 $\Omega$ which is quite close to the expected value. The DC gain of obtained input impedance (at $s = 0$) can be checked in a similar way for other type of converters.

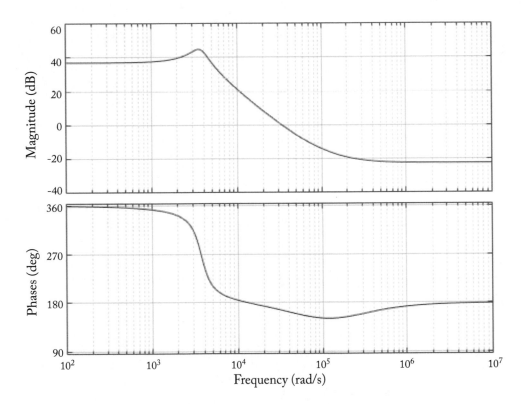

Figure 2.21: Control-to-output transfer function of the studied boost converter.

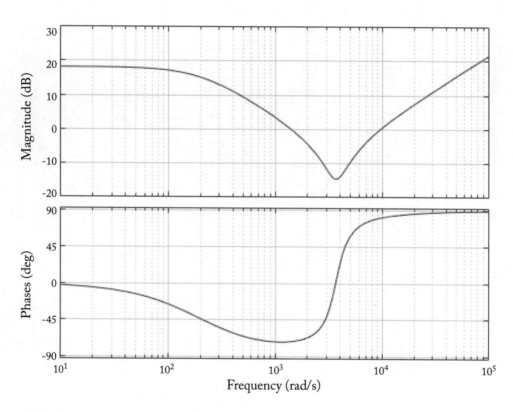

Figure 2.22: Open-loop input impedance of boost converter.

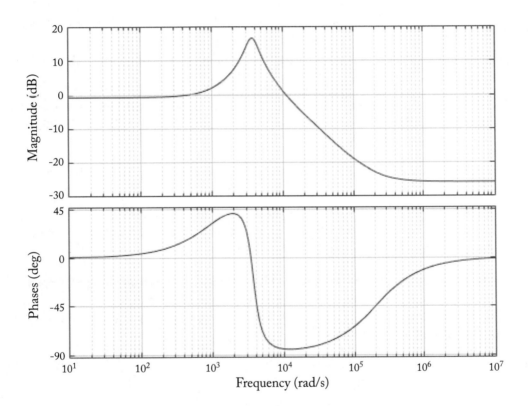

Figure 2.23: Open-loop output impedance of boost converter.

## 2.5 CONCLUSION

Input/output characteristics of DC-DC converters are important parameters. The input impedance helps the designer to select the suitable input source. The output impedance of the converter shows whether the converter can supply the output load successfully or not.

This chapter studied the input/output characteristics of buck, buck-boost, and boost converters. MATLAB® programming is used to do the mathematical machinery. Input/output characteristics of other types of converters can be extracted in a similar way shown in the chapter.

## REFERENCES

Voltage Regulator Module (VRM) and Enterprise Voltage Regulator-Down (EVRD), 11.1 Design Guidelines, Intel Corp., 2009. 39

Singh, R. P. and Khambadkone, A. M. A Buck-derived topology with improved step-down transient response, *IEEE Transactions on Power Electronics*, 23(6), pp. 2855–2866, 2008. DOI: 10.1109/tpel.2008.2005383. 39

Erickson, R. W. and Maksimovic, D. *Fundamentals of Power Electronics*, 2nd ed., Kluwer, Berlin, 2002. DOI: 10.1007/b100747. 40

Yao, K., Xu, M., Meng, Y., and Lee, F. C. Design considerations for VRM transient response based on the output imedance. *IEEE Transactions on Power Electronics*, 18(6), pp. 1270–1277, 2003. DOI: 10.1109/tpel.2003.818824. 40

Ahmadi, R., Paschedag, D., and Ferdowsi, M. Closed loop input and output impedances of DC-DC switching converters operating in voltage and current mode control. *IECON 36th Annual Conference IEEE Industrial Electronic Society 1*, pp. 2311–2316, 2010. DOI: 10.1109/iecon.2010.5675123. 40

Kazimierczuk, M. K. *Pulse-Width Modulated DC–DC Power Converters*, John Wiley, London, 2008. DOI: 10.1002/9780470694640. 40

Chen, D., Lee, F. C., and Chen, S. J. Evaluation of various adaptive voltage positioning (AVP) schemes for computer power sources. *Journal of Chinese Institution of Engineers*, 30(7), pp. 1137–1143, 2007. DOI: 10.1080/02533839.2007.9671341. 53

Lee, M., Huang, D. C., Chih-Wen, K., and Ben Tai, L. Modeling and design for a novel adaptive voltage positioning (AVP) scheme for multiphase VRM. *IEEE Transactions on Power Electronics*, 23(4), pp. 1733–1742, 2008. DOI: 10.1109/tpel.2008.924822. 53

Xiao, S., Qiu, W., Miller, G., Wu, T. X., and Batarseh, I. Adaptive modulation control for multiple-phase voltage regulators. *IEEE Transactions on Power Electronics*, 23(1), pp. 495–499, 2008. DOI: 10.1109/tpel.2007.912947. 53

Qahouq, J. A. and Arikatla, V. P. Power converter with digital sensorless adaptive voltage positioning control scheme. *IEEE Transactions on Industrial Electronics*, 58(9), pp. 4105–4116, 2011. DOI: 10.1109/tie.2010.2098366. 53

Singh, R. P. and Khambadkone, A. M. A Buck-derived topology with improved step-down transient response. *IEEE Transactions on Power Electronics*, 23(6), pp. 2855–2866, 2008. DOI: 10.1109/tpel.2008.2005383. 53

# Author's Biography

## FARZIN ASADI

**Farzin Asadi** received his B.Sc. in Electronics Engineering, his M.Sc. degree in Control Engineering, and his Ph.D. in Mechatronics Engineering. Currently, he is with the Department of Mechatronics Engineering at the Kocacli University, Kocaeli, Turkey.

Farzin has published 25 international papers and 6 books. He is on the editorial board of 6 scientific journals as well. His research interests include switching converters, control theory, robust control of power electronics converters, and robotics.

Printed in the United States
by Baker & Taylor Publisher Services